Electric Shocks: Are Drivers Rejecting the Electric Car?

Bob Williams

Electric Shocks: Are Drivers Rejecting the Electric Car?

History, Problems, Solutions and Alternatives

Springer

Bob Williams
CSi UK Ltd.
Abingdon, UK

ISBN 978-3-031-98819-6 ISBN 978-3-031-98820-2 (eBook)
https://doi.org/10.1007/978-3-031-98820-2

This Springer imprint is published by the registered company Springer Nature Switzerland AG
The registered company address is: Gewerbestrasse 11, 6330 Cham, Switzerland

Caveats

This work provides descriptions of many vehicles, past and present, but it does not attempt to provide a catalogue of BEVs (and would be quickly out of date if it attempted to do so). The vehicles cited are simply cited as examples of the point being discussed in the text.

The sales/production volumes/costing/pricing figures detailed in this work are approximations based often on incomplete uncorroborated diverse and multiple sources of information. And are provided to show trends of sales growth and sales patterns and should be interpreted solely within that context and the context of the objectives of this work. They should not and cannot be taken as precise sales or production data, and this work should not be cited as a reference source of such data.

Similarly, descriptions and images of vehicles, their motive power, battery size and range are provided to give a general description of the general nature and capabilities of a vehicle and are not to be read as a precise vehicle specification.

For precise data regarding exact production, sales, or model specifications the reader is referred to data released by the manufacturer. (Links provided in the Bibliography)

Particularly, the issue of assessing battery range is affected by conditions driven, speed, driving style, weather, congestion, age of battery, and other factors, not the least being several competing 'standard' ways to measure/assess range, and should be taken as a general indication only.

A comprehensive alphabetical bibliography of every name/model/person/regulation provides a link to where more information/source of information can be found at the end of this work.

This work uses the logical framework:

The four 'W's questions:

- Where are we?
- Why and how did we get here?
- Where are things heading?
- What is the real issue?

And the four 'F's resolutions:

- Find the problem(s).
- Face the problem(s).
- Find the solution(s) or alternatives.
- Futureproof the results.

Competing Interests The author has no competing interests to declare that are relevant to the content of this manuscript.

Contents

About the Author

Robert Kenneth Williams PhD, FCMA, CGMA (Bob Williams) is internationally recognized as one of the world's leading independent data capture consultants specializing in wireless technologies. He has more than 40 years of experience in Intelligent Transport Systems (ITS), Radio Frequency Identification (RFID), and Bar Coding. He has provided support to many of the world's leading companies and governments, has been the architect and editor of over 100 International Standards deliverables for ITS and RFID, and has authored several books on these subjects.

Abbreviations and Acronyms

AC	Alternating current
ACC	Automotive Cells Company (a joint venture between TotalEnergies and Stellantis)
ACEA	European Automobile Manufacturers' Association
ADAS	Advanced driver assistance system
AFOLU	Agriculture, Forestry, and Other Land Use
AGM	Annual general meeting
BEV	Battery electric vehicle
BMS	Battery management system
BVRLA	British Vehicle Rental and Leasing Association
bZ	Beyond Zero
CARB	California Air Resources Board
CATL	Contemporary Amperex Technology Co. Limited
CEVA	Chinese Electric Vehicle industry Association
CMA	Compact modular architecture
DC	Direct current
DMi	Dual mode intelligent
DRC	Democratic Republic of Congo
EC	European Commission
EEA	European Environment Agency
EGR	Exhaust gas regeneration
EMC	Electromagnetic compatibility
EPA	Environmental Protection Agency (USA)
EREV	Extended range electric vehicle
EUV	Electric utility vehicle
EV	Electric vehicle
FAW	First Automobile Works
FCA	Fiat Chrysler Automobiles
FCEV	Fuel cell electric vehicle
FSD	Full self-driving
GDP	Gross domestic product

GHG	Greenhouse gas
GM	General motors
GNSS	Global navigation satellite system
GOE	Great oxygenation event
GW	Gigawatt(s)
HFC	Hydrofluorocarbons
HICEV	Hydrogen internal combustion engine vehicle
hp	Horsepower
HREE	Heavy rare earth elements
HySE	Hydrogen small mobility and engine technology (Japan)
IARC	International Agency for Research on Cancer
ICCT	International Council on Clean Transportation
ICE	Internal combustion engines
ICEV	Internal combustion engine vehicle
IEEE	Institute of Electrical and Electronics Engineers
JiT	Just in time
JRG	Japan Railways Group
Kg	Kilogramme
km	Kilometre
kph	Kilometres per hour
kW	Kilowatt(s)
kWh	Kilowatt hour(s)
lbs	Pounds (Imperial weight)
LCC	London County Council
LCD	Liquid crystal display
LED	Light emitting diode
LFP	Lithium iron phosphate
LiAir	Lithium air
LiFePO	Lithium iron phosphate
Liion	Lithium ion
LiOH	Lithium hydroxide
LiS	Lithium sulphur
LPG	Liquid petroleum gas
LREE	Light rare earth element
LTH	Lund Faculty of Technical Engineering
MaaS	Mobility as a service
METI	Ministry of Economy, Trade and Industry (Japan)
MIIT	Ministry of Industry and Information Technology (China)
mpg	Miles per gallon
mph	Miles per hour
MPV	Multi-purpose vehicle
N·m	Newton-metre
NASA	National Aeronautics and Space Administration (USA)
NCM	Nickel cobalt manganese
NDC	Nationally Determined Contribution

NEC	Nippon Electric Company
NEV	New energy vehicle
NiCd	Nickel cadmium
NiMH	Nickel metal hydride
NOx	Nitrogen oxides and dioxides
OEM	Original equipment manufacturer
PFC	Perfluorocarbons
PHEV	Plug-in hybrid electric vehicle
PM	Particulate matter (e.g. PM2.5, PM10)
PMSM	Permanent magnet synchronous motor(s)
PPVUE	Pay-per-vehicle-usage of electricity
PV	Photovoltaic
R&D	Research and development
REE	Rare earth element
SAE	Society of Automotive Engineers
SAEVs	Electric and shared autonomous vehicles
SAIC	Shanghai Automotive Industry Corporation
SAV	Sports activity vehicle
SEK	Swedish Kroner
SMMT	Society of Motor Manufacturers and Traders (UK)
SPA	Scalable product architecture
SUV	Sports utility vehicle
TNO	(Netherlands Organization for Applied Scientific Research)
TWh	Terawatt hour
UN	United Nations
UN ECE	United Nations Economic Commission for Europe
UNFCCC	United Nations Framework Convention on Climate Change
USA	United States of America
USP	Unique Sales Proposition (USP)
V2X	Vehicle to anything (communications)
VAT	Value added tax
VLV	*Voiture Légère de Ville* (Light City Car)
VOC	Volatile organic compound(s)
ZEV	Zero-emissions vehicle

Chapter 1
The Need for the So-Called Green Solutions for Personal Transport

Abstract Chapter 1 summarises the need for battery electric vehicles (BEVs); the USP of the BEV; the differing requirements of urban and suburban lifestyles, needs for mobility of the workforce, effects of migration, and causes and consequences of social engineering.

Keywords Battery Electric Vehicles (BEVs) · Mobile workforce · Zero tailpipe emissions · Personal transport means · Global warming · Mobility-as-a-service · Mobility-on-demand

The Warming Planet

This book starts with the premise that the world is warming at a rate unacceptable to much of humanity and action is needed now to control/correct this as fast as humanity can do so. Battery Electric Vehicles (BEVs) are commonly seen as an important step to achieve such change.

The USP of the BEV

The benefits of using green electricity to power our vehicles, where and when achievable, is an obvious benefit in the fight to reduce the use of hydrocarbons, and its practicalities are the subject of this book. But, for the car salesman in the showroom, the Unique Sales Proposition (USP) of a BEV is this:

An internal combustion engine vehicle (ICEV) pays fuel duty and VAT or sales tax on every litre/gallon of fuel purchased—lots of it (over half of what you pay at the pump is tax in many countries). Green electricity is claimed to be now less expensive than that generated from hydrocarbons and bears only a reduced rate of VAT or no tax at all in some countries. Even better, is available at off-peak tariffs. So, as well as being good for the planet, offers the BEV driver a huge advantage in

B. Williams, *Electric Shocks: Are Drivers Rejecting the Electric Car?*,
https://doi.org/10.1007/978-3-031-98820-2_1

day-to-day running costs of the vehicle, which will quickly recover the purchase price/monthly lease payment difference between a BEV and an ICEV.

Taking the UK as an example, let us compare the recharging costs of a BEV with the fuel costs of an Internal Combustion Engine Vehicle (ICEV) per 100 km driven in the UK.

Let us start with the fuel costs of running a typical petrol drive ICEV. Forecourt service station petrol and diesel prices in the UK are around £1.50–£1.60 per litre at the time of writing. Let us be generous and say that a typical modern petrol-powered car might consume around 6–8 litres per 100 km (a car with 40–50 mpg fuel efficiency).

Say 7 litres per 100 km at an average of £1.55 = £10.85 per 100 km/£0.109 per km /£0.175 per mile.

By comparison, in the UK, off-peak electricity prices can range from £0.07 to £0.15 per kWh depending on the tariff. For this comparison, let us use an average off-peak rate of £0.10 per kWh. A typical BEV (Nissan Leaf, Tesla Model 3, etc.) consumes 15–20 kWh per 100 km. For our comparison, let us assume the BEV uses 18 kWh per 100 km.

Say 18 kWh per 100 km at an average of £0.10 per kWh = £1.80 per 100 km/£0.018 per km/£0.03 per mile.

These are rough comparisons of course. Petrol prices vary daily, electricity prices are revised quarterly and tariffs vary, mpg and kWh usage vary with the route and driver. But in rough terms, fuelling a BEV off-peak costs one sixth of the cost of fuelling an ICEV.

For the average family car, driving 12,000 km per year/1000 km per month that is an outlay of £18 per month/£216 per year as opposed to £109 per month/£1308 per year to fuel an ICEV.

Urban and Suburban Lifestyles

While environmentalists dream of city life where no-one uses a car but walks, uses (green) public transport or cycles, and that dream can be enforced on those living in the city centre, the reality for most in the developed world is that we have evolved our infrastructure into suburbs, towns, and villages interconnected by roads, and in many (but not all) cases with poor public transport options and a need to commute to work and travel to meet family, friends, socialise, and shop. Even in countries such as Singapore, or cities such as Paris or Oslo, where considerable investment has been made to provide (subsidised) public transport, or cities such as London or Singapore where they seek to tax the citizen out of private vehicle usage, there remains a high level of demand for personal transport.

In the less developed parts of the world, these concepts of comprehensive public transport options are the subject of an inconceivable unrealistic dreamworld where the reality is that the car or van—usually an ageing third-hand cast-off from a richer country—is the key to improving your life and income opportunity. Even in highly controlled states, the car has become the key to a better lifestyle. China is now by far the world's largest car market.

The conventional wisdom has become that currently cars are fuelled by hydrocarbons; hydrocarbon use contributes to global warming, electric vehicles have zero tailpipe emissions, so electric vehicles fuelled by green electricity offer the only viable solution.

A Dynamic Economy Needs a Mobile Workforce

Economists argue that successful economies require a mobile workforce, and this implies personal transport not only to commute to work but to stay in touch with family and friends (and thereby maintain the society), and this implies not only that the subject has access to personal transport means, but that the relatives and friends have such access as well ….i.e. access to personal transport means is a basic requirement for a successful economy.

A Rise in Migration

Whatever your personal view of inward migration, a volatile world and improved personal communications means have dramatically increased the migratory movement of people seeking safety and/or a better life. And whatever restrictions are enforced, global migration is a rising trend, and the evidence is that this implies the same transport pressures to those described above regarding a mobile workforce.

Social Engineering, Causes, and Consequences

I mentioned above, the environmentalists dream of city life where no-one uses a car but walks, uses (green) public transport or cycles, and city planners and environmentalists have ambition to impose that dream on those living in the city centre, indeed on all.

From the perspective of City Authorities and Planners, Governments, and the European Commission, even the United Nations. (UN ECE Workshop on Mobility as a Service: '……As more of the world's cities become congested and polluted, new business models and technologies are emerging to solve the mobility challenge. In 2014, global venture-capital investments into mobility services amounted to more than $5 billion, up from less than $10 million in 2009'.).

From this viewpoint, the so-called Mobility-as-a-Service (MaaS) which the European Commission. DG MOVE 'Smart Mobility and Services' website describes as: 'Mobility-as-a-Service (MaaS) will increasingly catalyse the public-private co-development and co-delivery of mobility and transport systems and services, as well as shared and open use of public space, data and infrastructure. Is potentially a way to remove traffic from city centres, reduce traffic congestion, reduce pollution, achieve CO_2 reduction targets, encourage healthier lifestyles, and improve quality of life for citizens, and even countering global warming'.

With MaaS, the argument goes, there is no need for personal private transport. This is a heavy social engineering paradigm, whose objectives are much more radical than simply improving the travel experience of travellers.

(UN ECE) '….to contribute to the global debate and facilitate consensus on the important role of the sector in sustainable urban mobility and transport development in the 2030 Sustainable Development Agenda………. At the same time urban passenger and freight transport have negative impacts through leading to congestion, pollution and to traffic safety challenges, to mention only a few most visible pressures. Increasingly, passenger and freight movements are intertwined in a zero-sum game, ……………… Concrete recommendations towards the development of sustainable public transport and urban mobility networks are provided considering that all public transport modes are efficiently interconnected and that cycling and walking are integral parts of such networks as well'.

'The principal prospects for decarbonisation are strong—better utilisation of underused assets in transport fleets and infrastructures can accommodate increasing demand and reduce the share of unsustainable travel modes. Smart mobility systems and services have the promise to contribute to the needed decarbonisation of the transport sector and might also help address persistent problems of congestion and accessibility. However, new innovations in technologies and use need to optimise the whole transport system not road-based car travel only to make a long-term contribution to decarbonisation'.

………'Technological, socio-demographic and behavioural change are facilitating a move towards multimodal transport—combining walking, cars, buses, bikes, trains and other forms of shared transportation. Driven by the transition from "owning" to "using", Mobility as a service (MaaS) enables multimodal mobility by providing user-centric information and travel services such as navigation, location, booking, payment and access that allow the use to consume mobility as a seamless service across all existing modes of transport.

Public and private business models, payment methods, technologies, and user choices will continue to coevolve alongside data sharing by users and public infrastructures, and increasing cooperation between the public and private sectors. Mobility as a service should also provide more cost-efficient mobility options to consumers and households by reducing vehicle acquisition and maintenance expenditures.

……….. The transformation and convergence of transport and mobility systems and services presents a unique opportunity to develop post-fossil, user-centric, smart mobility systems based on access to individual, public, shared and active mobility, rather than ownership of private automobiles.

………..Integrated Mobility-on-Demand services can contribute to modal shift to public transport and also address the spatial inefficiencies of private individual motorised transport. User-centric urban mobility systems will provide ubiquitous check-in/check-out user access to enable both inter- and multimodal mobility on demand and enhance overall transport efficiency. In future integrated and sustainable mobility-on-demand systems, electric mobility will become a component of both power and public transport infrastructure and systems. The smart integration of tariff structures, data and user interfaces as well as the disposition of rolling stock across these sectors is a central challenge, which requires new business models and scheduling, booking, navigating, ticketing and charging solutions'.

'.........Autonomous electric vehicles are expected to form a significant component of "mobility as a service" for urban transport. As with sharing models, autonomous vehicle (in public or private service) technology will blend with MaaS models and can potentially also enable ubiquitous smart traffic management. In deploying electric and shared autonomous vehicles (SAEVs) the benefits of these mobility strategies can be combined to greater effect'.

This is heavy social engineering and a world away from the more modest ambition of most MaaS service brokers. But pressures to reduce CO_2 emissions and reduce pollution will mean that there is continual and growing pressure to achieve these social ambitions. But whether or not that is in line with the purists' objective to 'improve the travellers journey experience' is another matter.

Driving your own car offers many advantages over using public transport. The three most important key advantages of travelling by your own car are usually cited as:

(A) It provides a single door-to-door service from A to B.
(B) 24/7 Availability.
(C) Faster travel.

But there are a host of other benefits, often taken for granted as well:

(D) You are in control of your own personal space.
(E) In your car, you have full control over your environment, including temperature, music, and seating, with no crowding or (assuming you can control your family) noise from other passengers.
(F) Driving allows you to carry more luggage, groceries, or other large items that would be cumbersome on public transport.
(G) Families with young children, pets, or those with mobility issues can find it easier to travel with the added space and convenience of a private car.
(H) You can travel to remote destinations, rural areas, or take spontaneous road trips that may not be accessible via public transport.
(I) The car allows you to travel directly from your home to your destination without transfers, waiting times, or walking to/from stations.
(J) You can choose your route, avoid congested areas, and take detours if necessary.
(K) You can take the most direct route, reducing travel time compared to indirect public transport routes.

No matter how much MaaS improves the public transport experience, it can't begin to overcome these advantages. That is why proponents of these modal shifts hung their hat on deploying electric and shared automated vehicles (SAEVs).

In this concept, you exchange monthly payments for owning/leasing your own car for a subscription to an automated driverless taxi service, of course a BEV (battery electric vehicle), where you call up the on-demand service on your mobile, when the car arrives you validate that it is you with your mobile or similar means, the car drives you to your destination and then sets off, driverless, to its next client. Suffice to say that the social engineers saw these services as an obvious and attractive way to remove the need to own a car—at least in a city, and solve pollution and parking and congestion problems, all in one go.

Well. It turns out, for the many reasons, that automated driving is much more complicated than teaching a car when to start and stop, turn left and right, and that although automated driver assistance is with us now, fully automated driving is still many years, if not decades away. Without fully automated driving, SAEV services are not possible.

Further, the provision of even very competent MaaS 'apps' and single payment journey planning/management, by itself, does little to prize drivers out of their vehicles. But too many planners/social engineers are relying on these MaaS Apps to change behaviour, just because the MaaS apps are created.

And it is now more than a decade since these lofty ideals were set, and not much progress has been made. Indeed, the prospect of shared automated electric vehicle services has receded into the future, along with the prospect of automated driving. It is true that public transport service information and update information services have improved markedly, especially via mobile phones, but the single seamless joined up MaaS provision remains the same pipedream as it was a decade ago. The only change being some cities resorting to removing private vehicles by onerous taxation and restrictive regulations.

Those who live in cities can usually commute to a fixed work location, or locations, by public transport, cycling or walking, but many need a car to visit relatives and friends, do the school run and ferry children around their busy social and educational calendars. Others need their car in order to access their workplace, or carry out their work: carers, maintenance workers, social workers, doctors, professional advisors, some delivery drivers, private car hire, many taxi drivers, etc. And then there are holidays, excursions, and out-of-town shopping trips. So, cars remain an essential requirement for most citizens.

Many, indeed most, workplaces have also located themselves in the suburbs to avoid city centre congestion issues, and so one (or more) car(s) are essential for most families. Having invested, by necessity, a significant amount of family income in a car or other type of vehicle, there has to be significant incentive or deterrent, to persuade the family to use another transport means for all travel, except perhaps for travel to/from the city centre.

This is partly because the family has only to meet the additional 'marginal' cost of using the already purchased vehicle, and partly because (and this should not be underestimated) it represents the most convenient and comfortable means to achieve the journey, and for most journeys outside of the city centre, 'Mobility as a Service' multimodal journeys are a less attractive option.

The willingness to change travel behaviour is very questionable. Outside of city central areas, why would a car driving traveller who can easily, and largely comfortably drive his car from start point (a), seamlessly to end point (d), and back again, elect to choose to travel from (a) to (b), wait for a connection, possibly in inclement conditions, to catch a link to (c) in order to wait for another connection possibly in inclement conditions to catch a link to (d), and then do the same thing to return later?

However, as younger people generally have to struggle harder than their forebears to both own a vehicle, and rent or buy a home, and many also have student loan repayments to meet, and many are concerned enough about changing climatic conditions, they may be more inclined to test innovative travel modalities.

That said, there is limited evidence to support this. The Economist (February 15th 2020) reported 'Everybody wants buses. They just do not want to use them' and detailed that Transport for London reported that the average 17–24 year old took 2.3 public transport per day in 2011–2012, but only 1.7 in 2018–2019. Whereas the bottom fifth poorest without access to a car fell from 49% to 46% in the same period, and the second fifth poorest group without access to a car fell from 36% to 28% in the same period. Among the elderly, free bus pass movements also declined by around 20% despite an increase in those qualified for free bus passes.

For many therefore, the car remains an essential item, and the challenge is to stop it polluting, make it 'greener'. And the obvious answer for many, including most regulators, is to switch to the battery-powered electric vehicle (BEV).

Chapter 2
The History of Electric Vehicles

Abstract Chapter 2 provides a concise but detailed history of the evolution of transport means and the role of electric vehicles in that evolution. It summarises the development of early electric vehicles, the challenge of the internal combustion engine, the evolution of specialised electric vehicles such as trams, trolleybuses, golf carts, and milk floats. It presents the intrinsic problem of fuel storage for electric vehicles and summarises the demise of the electric car by the 1930s.

Keywords Fixed infrastructure systems · Rechargeable battery · Lead–acid · Electric vehicle · Internal combustion engine (ICE) · Limited range · BEV

Early Evolution of Transport Means

From time immemorial, through to the advent of canals in the eighteenth century; through the early stages of the industrial revolution, and even today in less developed parts of the world, the prime means of moving people and goods around has been the quadruped—the horse, donkey, camel, dog, etc.—and teams of them yoked together.

The infrastructure required for these transport systems is minimal because the quadruped can carry its own food as part of its load and needs only access to water, or a means of collecting water, in order to be fit for purpose (although over time in socio-civilised countries structures of smithies, coaching houses, inns and veterinary support evolved through business enterprise and opportunities). The quadruped does not require a sophisticated road, rail, or waterway, just reasonable accessibility. The required investment is minimal—you only have to inherit, breed, or buy a quadruped and keep it fed.

The advent of the steam locomotive was made by Richard Trevithick, a British mechanical engineer and inventor, who successfully harnessed high-pressure steam and developed the first working railway steam locomotive. The world's first locomotive-hauled railway journey took place on 21 February 1804, when Trevithick's unnamed steam locomotive hauled a train along the tramway of the

B. Williams, *Electric Shocks: Are Drivers Rejecting the Electric Car?*,
https://doi.org/10.1007/978-3-031-98820-2_2

Penydarren Ironworks, in Merthyr Tydfil, Wales. He also constructed what is widely recognised as the world's first steam powered horseless carriage which he demonstrated in London in 1803. These concepts were quickly taken up by Matthew Murray, John Blenkinsop, William Hedley, Timothy Hackworth, and the Stephenson's (largely to move coal from the mine to ports and factories and cities). In 1805, Trevithick adapted his high-pressure engine to propel a barge with the aid of paddle wheels, while George Stephenson's son Robert's company became the first to carry passengers on a public railway in 1825.

These developments led to the wide spread and development of railway systems and infrastructure in Great Britain, the USA, and Europe principally in the period 1840–1900, but have continued to evolve into the high speed rail and metropolitan train systems (now largely electric powered) that now encompass the world.

But railway systems and canal systems are fundamentally different from quadruped systems in that their architecture is fixed and has to be in place and functioning BEFORE any service can be provided. And that means that the system architect chooses, or is instructed, as to where the stopping stations are (whereas the driver of the quadruped-based system can make on the spot dynamic decisions as to where and when to stop). In theory, both types of system can determine how long to stop for, but a railway system is a fixed and finite resource that usually requires multi-unit occupancy to maximise financial viability, which will usually restrict options available, and, if steam powered, may also require time for refuelling and/or taking on water.

Fixed infrastructure systems such as canals and railways therefore require very heavy upfront investment to create the infrastructure, societal demand (or at least societal consent), and, especially in the case of railways, significant ongoing costs to maintain and operate the infrastructure, BEFORE any service can operate.

Fixed infrastructure systems also affect the development of, and restrict the mobility of, societies, in that associated infrastructure developments are attracted to and focused around the predetermined station stops. This can be beneficial for social engineering and its long-term significance is often undervalued against short-term viability.

Although there are a few notable exceptions, very few railway systems operate without some form of direct or indirect subsidy, or both. Rail subsidies vary in both size and distribution, with some countries funding the infrastructure and others funding trains and their operators, while others have a mixture of both. Rail subsidies are largest in China, Europe, and India, while the USA has relatively small subsidies for passenger rail with freight not subsidised at all. The Japan Railways Group (JRG) famously operates main line services without subsidy, but when it was created out of the loss making Japan National Railway in 1987, much of the investment accumulated losses were written off. A growing number of local train services in Japan are operated by the so-called Third Sector public/private partnerships that receive subsidies.

Early Horseless Carriages

As we have seen in the text above, in 1803, what is widely said to have been the first horseless carriage was a steam-driven vehicle demonstrated in London, England, by Richard Trevithick.

But French engineer Nicolas-Joseph Cugnot had had already demonstrated a steam powered vehicle in 1769. An army captain, Cugnot was one of the first to successfully employ a device for converting the reciprocating motion of a steam piston into a rotary motion by means of a ratchet arrangement. A small version of his three-wheeled 'fardier à vapeur' (steam dray) was made and used in 1769 (a fardier was a massively built two-wheeled horse-drawn cart for transporting very heavy equipment, such as cannon barrels).

Cugnot's 1770 fardier à vapeur, is preserved at the Musée des Arts et Métiers, Paris. According to the records of the Musée, this full-size version of the fardier à vapeur was built, specified to be able to carry four tons and cover two lieue (7.8 km, or 4.8 miles) in 1 h (a performance it never achieved in practice). The vehicle weighed about 2.5 tonnes and had two wheels at the rear and one in the front where the horses would normally have been. The front wheel supported a steam boiler and driving mechanism. The power unit was articulated to the 'trailer' and was steered from there by means of a double handle arrangement. The vehicle was reported to have been very unstable owing to poor weight distribution, and further development appears not to have continued.

William Murdoch developed Britain's first working model of a steam engine carriage, or locomotive, shortly after, in 1781.

A colleague, Thomas Wilson, wrote to James Watt on Murdoch's 'new scheme': *It is no less than drawing carriages upon the road with steam engines...he says that what he proposes, is different from anything you ever thought of, and that he is positively certain of its answering and that there is a great deal of money to be made by it.*

Accounts exist from witnesses who 'saw the model steam carriage run around Murdoch's living room in Redruth' in 1784.

Murdoch's working model was a three-wheeled vehicle about a foot in height with the engine and boiler placed between the two larger back wheels with a spirit lamp underneath to heat the water and a tiller at the front turning the smaller front wheel.

Step forward to the 1820s, a surgeon called Goldsworthy Gurney built steam-powered road vehicles. By 1825, he had started practical work on a steam carriage, filing a first patent for 'An apparatus for propelling carriages on common roads or railways—without the aid of horses, with sufficient speed for the carriage of passengers and goods'. His work encompassed the development of the blastpipe, which used steam to increase the flow of air through a steam engine's chimney, so increasing the draw of air over the fire much increasing the power-to-weight ratio of the steam engine. But he ran out of funding and the enterprise failed.

In the USA, in 1865, Massachusetts machinist Sylvester Roper built at least seven steam carriages and two steam motorcycles. They weren't considered practical vehicles but became popular attractions at circuses and fairs.

Benjamin Franklin

Benjamin Franklin (1706–1790) predates steam power. Franklin was not only a scientist, but also an engineer. Even before James Watt invented his rotary motion steam engine, Benjamin Franklin had devised a working electric motor.

Electrical technology in Franklin's day did not of course have the benefit of a supply of 'mains' electricity. Research was focused on the capture of and means of storage of what became called the 'sublime fluid', developing crude electrostatic generators and an early form of the capacitor, which they called the 'Leyden jar'. Scientists employed these new devices to explore how electricity could be generated, stored, and transmitted, but harnessed too small amounts to find any practical uses.

Franklin sought to change that.

Franklin's Electric Motor

In 1748, Franklin invented what he called the 'electric wheel'.

This device comprised of a vertical shaft that was free to rotate, from which several glass bars extended like spokes, much like a wheel. Each bar was tipped with a small brass cap. Placing the terminal of a 'negatively charged' Leyden jar near the 'wheel' allowed the brass capped glass assemblies to act as capacitors; and as each assembly charged up, it tended to be repelled from the Leyden jar.

A second, 'positively charged' Leyden jar was placed nearby and had the opposite effect, attracting the brass capped glass assemblies towards it. As a result, the 'wheel assemblies' began to rotate, and would continue to do so until the power in the charges in the Leyden jars was dissipated.

The motor was only strong enough to continue turning at 12–15 rpm when carrying a small load so could not compete with water wheels as a practical source of power. And electricity remained a non-viable option until a reliable means to store or transmit significantly useful amounts of it were developed.

Early Electric Vehicles

In 1828, a Hungarian priest and physicist Ányos Jedlik invented an early type of electric motor and created a small model car powered by his new motor.

Between 1832 and 1839, Scottish inventor Robert Anderson invented a crude electric carriage.

In 1835, Professor Sibrandus Stratingh of Groningen, the Netherlands and his assistant Christopher Becker from Germany also created a small-scale electric car, powered by non-rechargeable primary cells.

In 1860, the Frenchman Gaston Planté invented the first practical version of a rechargeable battery (based on lead–acid chemistry), which it turned out was not only to provide power for electric carriages, but finished up providing the electricity required for hydrocarbon fuelled vehicles as well. Combined with the development of the Gramme machine (also known as a Gramme ring, Gramme magneto, or Gramme dynamo) in 1871 by Zénobe Gramme, a Belgian inventor, it provided the electricity required and recharged. The Gramme machine is an electrical generator that produces direct current and was built as either a dynamo or a magneto.

What is likely the first human-carrying electric vehicle with its own power source was tested along a Paris street in April 1881 by French inventor Gustave Trouvé. In 1880, Trouvé improved the efficiency of a small electric motor developed by Siemens and using the recently developed rechargeable battery, fitted it to an English James Starley tricycle, so inventing the world's first electric vehicle. Although this was successfully tested on 19 April 1881 along the Rue Valois in central Paris, he was unable to patent it.

The first electric car in the USA was developed in 1890–91 by Scottish-born William Morrison of Des Moines, Iowa; the vehicle was a six-passenger wagon capable of reaching a speed of 23 kmh (14 mph).

Practical, commercially available electric vehicles appeared during the 1890s. Electric battery-powered taxis became available at the end of the nineteenth century. In London, Walter Bersey designed a fleet of such cabs and introduced them to the streets of London in 1897.

At the same time, in New York City, the Samuel's Electric Carriage and Wagon Company began running electric hansom cabs. By 1898, with up to 62 cabs operating, it was reformed as the Electric Vehicle Company.

Electric vehicles had a number of advantages over their early-1900s competitors. They were popular because they did not require a manual effort to start (didn't need a hand crank to start the engine), nor the long start-up times associated with steam engines. They did not have the vibration, smell, and noise associated with gasoline cars.

Electric cars found popularity among affluent customers who used them as city cars, where their limited range proved to be no disadvantage. Henry Ford's wife Clara famously drove a Detroit electric car in the 1910s (source: SAE).

According to Wikipedia:

> Acceptance of electric cars was initially hampered by a lack of power infrastructure. In the USA by the turn of the century, 40% of automobiles were powered by steam, 38% by electricity, and 22% by petrol. A total of 33,842 electric cars were registered in the United States, and the USA became the country where electric cars had gained the most acceptance. Most early electric vehicles were massive, ornate carriages designed for the upper-class customers that made them popular. They featured luxurious interiors and were replete

with expensive materials. Electric vehicles were often marketed as luxury cars for women, which may have generated a stigma among male consumers. Sales of electric cars peaked in the early 1910s. There were over 300 listed manufacturers who produced a vehicle in the USA until 1942.

In 1910s, the 'Standard Electric' used Westinghouse electric motors and claimed to have a range of 110 miles on a charge. It was operated by a tiller from the left-hand side. The controller had six forward speeds and had a top speed of 20 mph. The model M was a closed model Coupe or open Runabout and priced from $1785 to $1900.

The Beardsley Electric Company, founded in Los Angeles by Volney S. Beardsley in 1913, produced 661 electric cars between 1914 and 1917. A typical cruising range was 115 km (70 miles). One model, the 'light town car', in 1916 was priced at US$1285 (equivalent to $35,980 in 2023).

Among the most notable of the records achieved by pre-1900 electric vehicles was the breaking of the 100 kmh (62 mph) speed barrier, by Camille Jenatzy on 29 April 1899 in his 'rocket-shaped' vehicle '*Jamais Contente*', which reached a top speed of 105.88 kmh (65.79 mph). An electric vehicle held the vehicular land speed record until around 1900.

The Challenge of the Internal Combustion Engine

The internal combustion engine (ICE) became feasible in 1876 when Nikolaus Otto, a German engineer, developed the first four-stroke internal combustion engine. Known as the 'Otto engine', it used a mixture of fuel and air, ignited by a spark plug. Otto's engine introduced the four-stroke cycle—intake, compression, power, and exhaust—that remains the foundation for modern gasoline engines.

Nikolaus Otto's engine revolutionised the industrial sector, but it was a German inventor, Karl Benz, who applied the internal combustion engine to transportation. In 1885, Benz developed a gasoline-powered automobile, which incorporated Otto's engine. This invention is widely regarded as the birth of the internal combustion engine powered automobile.

The late nineteenth and early part of the twentieth century saw significant advancements in internal combustion engine technology. In 1892, Rudolf Diesel, a (Paris born) German inventor, introduced the diesel engine. In 1902, the development of the carburettor further improved engine efficiency. The invention of the electric starter by Charles Kettering in 1912 eliminated the need of a hand starting crank. The invention of the muffler by Milton O. Reeves and Marshall T. Reeves in 1897 significantly reduced the noise to tolerable levels. Finally, the initiation of mass production of internal combustion powered vehicles by Henry Ford brought their price down.

By contrast, the price of similar electric vehicles continued to rise; by 1912, an electric car sold for almost double the price of a gasoline car, and the ease of refuelling meant that the operating range of the internal combustion engine vehicle became almost limitless, which better suited the increased travel ambition of citizens all around the world.

Most electric car makers ceased production at some point between 1910 and 1920.

Electric vehicles became popular for applications where they could be connected to mains supply of electricity (trains, trams, and trolley buses) or their limited range did not pose major problems (forklift trucks, milk floats, and golf carts). But by the 1920s, the early heyday of electric cars had passed, and a decade later, the electric automobile industry had effectively disappeared.

Solutions Encouraged by the Consequences of WWII

The Second World War disrupted the supply of both petrol and diesel to domestic markets in Europe in two different ways regardless of which side you were on. Firstly, armed naval conflict both sunk ships and impeded ship movements, which reduced the import of oil. Army front lines disrupted both pipeline and tanker supplies of all oil product. Then the priority for the limited amount of product that was available was directed to the armed forces to fuel tanks, airplanes, jeeps, personnel carriers, and the large fleets of vehicles needed in logistical support of the armed forces.

Fuel was strictly rationed and many vehicles were converted to use wood gas in their engines in addition to the normal but rationed gasoline. But in many cases wood was also a scarce commodity. Ethyl alcohol and 'syngas' (coal gasification) were developed for industrial use and experimentally used in vehicles, but not on a wide scale.

In France Pierre Faure, Citroen and Peugeot experimented with electric cars, but only Peugeot went into 'mass' production with the 'VLV'. The *Voiture Légère de Ville* (Light City Car) was powered by four 12 V batteries giving it a claimed top speed of 36 km/h (22 mph) and a range of 50 miles (80 km). According to *Auto Express*, only 377 were built. Faure launched his micro coupe in 1946, but it was doomed to commercial failure as soon as internal combustion vehicles became available again following the recently ended war.

In the USA, with its own domestic supply of gasoline, and not being an active theatre of war, evolution of the internal combustion engine vehicle continued apace, indeed increased due to the prosperity the additional industrial activity induced by the war.

But the weakness of the business case for electric vehicles was such that neither side expended much effort to attempt to make it a viable option. Neither German nor British auto manufacturers produced electric models because their factories were directed to support the war effort.

However, postwar oil shortages in Japan resulted in the successful production and sale of the 'Tama', a 2-door, 4-seat electric car launched in 1947. A government performance test calculated that Tama could run a maximum of 96 km on one battery recharge at speeds up to 35 kph. Tama production was directed primarily to supply taxis to urban areas. Some records indicate Tama and other EVs accounted for at least 3% of all automobiles in Japan in 1949, (this was a time of very limited car sales because of the postwar recession, and a low vehicle population because of destruction during the war).

Trams, Trolley-Buses, and Milk Floats

But despite the failure of the business model for electric cars, electric powered vehicles succeeded in markets where range and refuelling were not a problem, or there was a constant supply of mains electricity.

In 1881, the first commercially successful electric tramway was built by Werner von Siemens in Lichterfelde near Berlin. The arrival of the electric tram on London's streets in the early 1900s brought cheap transport to the masses, significantly replacing 'messy' horse drawn buses.

The first tram service in London opened in 1901. Trams could carry twice as many people as horse drawn or motor buses, and in greater comfort. They were cheap to run, so fares were low, and they were quick and frequent. Despite competition from the first motor buses, the number of passengers using trams grew. The London County Council (LCC) was committed to a policy of low fares to encourage people to use trams and other tram operators were forced to follow suit. At their peak, over 3000 trams carried a billion passengers a year over 366 miles of track. (source London Transport Museum).

After the First World War, the tram began to lose its dominance as the capacity and comfort of the motor bus improved and competed for passengers. In 1923, buses carried more passengers than the trams for the first time, and most of London's tramways were running at a loss.

LCC had adopted the 'conduit' system of supplying power to trams by means of a live rail buried under the road surface, rather than the unsightly, if cheaper, overhead wires. The conduit track was expensive to build and maintain. By the late 1920s, the pre-war trams were spartan and in need of modernisation. The new buses offered higher standards of comfort, were more flexible to route and drive, and offered a lower-cost option. Many tram operators were soon converting routes to buses or trolley buses. Trolley buses, like trams, are powered by electricity taken from overhead wires, but run on pneumatic tyres, not on rails.

The first rail-less electric trolley vehicle was demonstrated in London in 1909, and 2 years later the first trolley bus services were started in Leeds and Bradford. London's first trolley bus service was started in 1931, replacing a tram route.

The trolley bus was cheap to run and more manoeuvrable than the tram. Existing equipment could be easily converted at half the cost of modernising the tram system. The trolley buses were popular. They were fast, clean, and quiet. At the peak of its service in 1952, following the demise of the tram, London's trolley bus fleet was the largest in the world, with 1811 vehicles on 254 miles of route.

Increasingly, trams were seen as noisy and dangerous to road users, and costly to taxpayers. In 1931, a Royal Commission had recommended replacing trams with trolley buses, and the conversion programme began in 1935. By 1940, half of London's trams had been scrapped. (source: London Transport Museum).

But diesel powered buses were cheaper still and were not limited to a rail or overhead power network, were flexible and able to quickly service expanding suburbs, and even better, did not require upfront investment in, nor ongoing maintenance costs of, a rail or overhead power network. The replacement programme began in 1959, and London's last trolley bus ran on 9 May 1962. After more than 60 years, electric street transport in London was at an end.

Mainland Europe however has maintained its love affair with the tram, whose networks have continued to expand. Mostly powered by overhead cables, whose installation and maintenance costs are lower than the 'conduit' system used in London, tramway systems grace the streets of most cities in Europe and are seen as part of the 'green' emission-free solution to city transportation.

Although it operates several tramway networks, China has maintained its love of the trolley bus. Trolley buses have provided regular public transport service in 27 different cities in China at one time or another. Currently, 11 urban systems are in operation, including Beijing, Shanghai, Guangzhou, Wuhan, Qingdao, and Jinan, among other locations. Beijing's trolley bus system is the most extensive in China, boasting 31 routes and a fleet of over 1250 dual-mode single and articulated trolley buses. Interestingly, Shanghai's system holds the distinction of being the world's oldest continuously operating trolley bus system, having been in operation since November 1914. (Source Copilot, 15/07/2024).

The USA and Canada operate tramway networks (which they more commonly call 'streetcars') in several cities. Some are historical and remain as part of their tourism attraction; others are modern streetcar/tramway systems. Currently, only Toronto operates a streetcar network that remains largely unchanged in layout and mode of operation. Some older surviving lines and systems in cities such as Boston, Cleveland, Mexico City, Newark, Philadelphia, Pittsburgh, and San Francisco were infrastructure-heavy, featuring tunnels, dedicated right-of-way, and long travel distances. Additionally, cities such as Portland, Seattle, Salt Lake City, Tucson, Oklahoma City, and Atlanta have built new modern streetcar lines or combined modern light rail and streetcar systems. (Source: Copilot, 15/07/2024).

Only five American cities have trolley buses: Boston, Dayton, Philadelphia, San Francisco, and Seattle.

A milk float is a vehicle specifically designed for the delivery of fresh milk. Today, milk floats are usually battery electric vehicles (BEV), but they were formerly horse-drawn floats. They were once common in many European countries, particularly the UK, and were/are still operated by local dairies. Operating over a limited rage, at slow speed, and usually operated only part of the day, they are well suited to the constraints of the electric vehicle. However, in recent years, as the number of supermarkets, small independent grocers and petrol stations, and convenience stores stocking fresh milk has increased, and people now have refrigerators, most people have switched from regular daily home delivery to obtaining fresh milk from these other sources and so the market for milk floats has reduced significantly.

Other local delivery services have adopted the electric vehicle. Famously, the upmarket London retailer Harrods began using electric vans for deliveries in 1919. These American-built Walker electric vans became a familiar sight on the streets of London, serving the famous department store until the 1930s when Harrods renewed its fleet. The 1939 Harrods electric van had a range of 60 miles per recharge and a top speed of 18 mph. After a number of years without using electric vehicles, the store has recently returned to using them (with modern Nissan electric vehicles).

Golf carts are another purposed vehicle well suited to the constraints of the electric vehicle. The first electric golf cart was custom-made in 1932. From the 1930s until the 1950s the most widespread use of golf carts was for those with disabilities who could not walk far. But by the mid-1950s, the golf cart had gained wide acceptance with US golfers, and their use has subsequently spread rapidly around the world both for golf and as local maintenance support vehicles.

The Challenges of the Fuel Storage

This work assumes the claims that electricity supplied from the network can and will become green. The author is aware that there are challenges to some of the claims being made concerning the so-called green; 'zero emission' energy, but this work is the wrong place to debate such issues. This work is about the viability of the current battery electric vehicle and will stay focused on these issues. The issue in this section is about storage of electricity on board the vehicle.

We have seen from the historical summaries above, that from the days of Benjamin Franklin (and even before) that capture, and particularly storage of the 'sublime fluid' posed, and still poses, its greatest challenge. If you are using water as a power source mechanism you simply store it in a jar, bucket, cistern, or reservoir until you need it. Similarly with gasoline (though that has the complication of flammability). If you are using wood, peat, or coal, you simply pile it up in a heap. If you are using gas, you compress it into a tank. Not so electricity. It is generated and dissipates. Its 'energy' is not per se a store (like a heap of logs), but a transient effect of the movement of electrons.

Britannica describes electricity as 'Electricity, (is a) phenomenon associated with stationary or moving electric charges. Electric charge is a fundamental property of matter and is borne by elementary particles. In electricity the particle involved is the electron, which carries a charge designated, by convention, as negative. Thus, the various manifestations of electricity are the result of the accumulation or motion of numbers of electrons.'

And describes an electric charge as:

'basic property of matter carried by some elementary particles that governs how the particles are affected by an electric or magnetic field'.

As soon as wires are connected to the battery, completing the circuit, ions from the high-energy electrode (the positive terminal) move through the electrolyte solution towards the low-energy electrode (the negative terminal). At the same time, electrons also move from negative to positive through the wires. This controlled movement of charged particles is the electric current.

So if you are not connected to a supply of mains electricity and you can't store the electricity (like a sack of coal or a pile of wood), what you need to do is store the components of a chemical reaction that excites its electrons thus producing an electric charge. We commonly call such a device a battery.

A battery is an assembly of two or more galvanic cells capable of energy conversion into the provision of an electric charge (motions of electrons) when a circuit between the negative and positive electrodes is made. Every battery (or cell) has a cathode, or negative plate, and an anode, or positive plate. These electrodes have to be separated by and are often immersed in an electrolyte that permits the passage of ions between the electrodes.

The amount of electrical energy produced depends on the reaction between the chemicals, as does the length of time the reaction continues to generate electricity (plus a host of other factors).

Then you encounter the issue that every charge is finite, and what do you do when it dissipates? Throw it away and get another, or find a way to recharge the battery.

A primary battery or primary cell is a battery (a galvanic cell) that is designed to be used once and discarded, and it is not rechargeable unlike a secondary cell (rechargeable battery). In general, the electrochemical reaction occurring in the cell is not reversible, rendering the cell unrechargeable. As a primary cell is used, chemical reactions in the battery use up the chemicals that generate the power; when they are gone, the battery stops producing electricity. In contrast, in a secondary cell, the reaction can be reversed by running a current into the cell with a battery charger to recharge it, regenerating the chemical reactants. Primary cells are made in a range of standard sizes to power small household appliances such as flashlights and portable radios. (And still constitute more than 90% of the battery market).

When you recharge a secondary cell battery, you are essentially reversing the process of discharging it. During discharge, the battery releases energy as chemical reactions occur between its electrodes and electrolytes. When you recharge it, you force current to flow in the opposite direction, restoring the battery's chemical state. (e.g., in a lead-acid battery, when recharging, the positive terminal of the direct current source connects to the battery's positive terminal (anode), and the negative terminal connects to the negative terminal (cathode). Hydrogen ions move towards the cathode, and sulphate ions (SO_4^{2-}) move towards the anode. At the anode, SO_4^{2-} combines with water to form sulphuric acid (H_2SO_4) and oxygen (O). Oxygen reacts with the anode material ($PbSO_4$), converting it back to lead dioxide (PbO_2). At the cathode, hydrogen ions take electrons from it, forming hydrogen gas (H_2). The hydrogen gas reacts with the cathode material ($PbSO_4$), converting it back to lead (Pb), thus 'recharging' the battery's ability to react and generate the electricity. In a lithium-ion battery, positively charged lithium ions move from the cathode to the anode during charging, while electrons concentrate on the anode. So, recharging replenishes the battery's stored energy by resetting its chemical composition.

The type of chemical reaction chosen for the battery therefore not only affects the amount of electrical energy produced, but the duration of its production, time to recharge, size, and weight of the battery (and lifetime, charging curve, temperature gradient, etc.).

As we have seen above, the electric vehicle only became feasible when then rechargeable lead-acid battery was developed. And, as we will see below, the modern BEV is dependent on more efficient battery technologies.

Chapter 3
Enter the BEV, the 'Solution' to the Politicians and Environmentalists Prayers

Abstract Chapter 3 describes the entry and detailed evolution of the BEV. It provides a summary of every significant BEV development in the period 1950–2000 (including approximate sales volumes per model/year), the Kyoto Protocol, the initiatives of the new millennium and the growth and influence of the Chinese market, providing a summary and evolution of every major Chinese automotive company (including models and approximate sales volumes per model/year) and details of Chinese state development programmes. It then describes the 2010–2020 entry of Nissan, Tesla, Volvo, and the big established auto manufacturers, together with descriptions of all of their BEV models, approximate prices ranges, driving range, and approximate sales volumes by model. The chapter goes on to describe the relevant provisions of the 2015 Paris Agreement and IPCC Reports and Climate Action. In that context, it continues to assess changes in land use and urbanisation, natural factors, oceanic emissions and changes, cyanobacteria, sources of human-generated CO_2 emissions, the number of cars or number of humans, and the Darwinian outcomes.

Keywords Battery Electric Vehicle (BEV) · Extended Range Electric Vehicle (EREV) · Intergovernmental Panel on Climate Change (IPCC) · Kyoto Protocol · Hybrid · 2015 Paris Agreement (COP21) · Greenhouse gases · Energy Sector

Early Modern BEVs

So how did the battery electric vehicle (BEV), consigned to the museum by the 1930s, escape from the museum of discarded technologies?

We have seen above the failure of Pierre Faure's coupe and Peugeot's 'VLV'—The *Voiture Légère de Ville*—as soon as a distorted market was normalized and internal combustion vehicles became available again following the recently ended war. And we have seen the brief success of the 'Tama' in 1947 in Japan while there was a chronic shortage of gasoline.

B. Williams, *Electric Shocks: Are Drivers Rejecting the Electric Car?*,
https://doi.org/10.1007/978-3-031-98820-2_3

The Henney Kilowatt was an electric car introduced in the USA for the 1959 model year. The car was modelled using body parts of the Renault Dauphine. Only 47 cars were sold over the two model years, mostly to electrical utility companies.

In the 1960s, incited by clean air acts in the USA and Europe, General Motors, American Motors Corporation, Enfield, Scottish Aviation experimented with electric vehicles, but none reached production.

GM's 1966 Electrovair and more particularly the Electrovair II were typical developments (based on the Corvair, its lightest model), pushing new technologies to meet clean air demands, but failing to produce a commercially viable product. Electrovair II used a silver zinc battery which delivered high peak power and good energy storage while only adding 360 kg of additional weight (a conventional lead-acid battery would have added 1180 more kilogrammes more than doubling the vehicles weight). But Electrovair II never made it past the experimental phase. The car's battery provided between 40 and 80 miles of range, and the battery wore out after 100 recharges.

Although AMC experimented with nickel-cadmium, and latterly lithium batteries from Gulton (maker of 'Exide' batteries) the futuristic looking AMC 'Amitron' included many advanced features, including regenerative braking and advanced battery designs, to provide a 150-mile (240 km) range on a single charge. Development ended because of technology issues and the high cost of batteries. It was later relaunched as the 'Electron' in 1977, but faced the same battery cost issues.

The Enfield 465 was a small 2 + 2-seater electric car built only in prototype form in 1969. It was equipped with a 48 V, 4.65 bhp (3 kW) electric motor and had a plastic body and failed crash testing. It evolved into the Enfield 8000 (Electric City Car) that had 8 bhp (6 kW) motor and aluminium body. 120 Enfield 8000s were built most were used by the Electricity Council and southern English electricity boards and the venture failed to find commercial success.

Designed as the second car in a household, Scottish Aviation's 'Scamp' would probably be used by only one member of that household to undertake, probably on a daily basis, a specific, known journey or pattern of journeys, such as the 'station run' or trip to the supermarket—after which it would be at home and could be completely recharged overnight from a standard 13 amp domestic socket. It was based on airport baggage trucks. A top speed of 30 mph, a range of 25 miles (40 km) and cheap. But emphasis on that latter issue resulted in a fragile poor construction that fell apart during endurance testing and the project was cancelled.

Whether as a result of publicity about the Scamp, or coincidentally, in 1967 Ford UK developed a small city commuter car, 'the Comuta' aimed at the British market, by Ford's Dunton Technical Centre. It used four 12-volt lead-acid batteries, had a range of 60 km (37 miles) at a speed of 40 km per hour (25 mph).

In 1967, John Gould started Unique Mobility. The company intended to be a car manufacturer, but they would make only one car. In 1979, Unique Mobility launched the 'Electrek'. The Electrek has a 32 horsepower General Electrics motor connected to a Volkswagen Fox 4 speed manual transmission and featured regenerative braking. Top achievable speed was about 65 mph, range about 100 miles. The car was powered by 16 6-volt batteries wired in series which ran into a custom motor controller designed specifically for the Electrek. Less than 100 were made.

In eastern Europe, Czechoslovakia in 1969–70, produced the EMA1, a compact electric car developed by the Research Institute of Electric Machines in Brno, which used separate motors on two wheels; top speed 31 mph and a range of 31 miles. It was designed for two adults and two children. It never made it to production.

BMW developed its first electric car based its 1602 model. The 'BMW 1602 Elektro-Antrieb' was first unveiled at the 1972 Olympic Games in Munich, Germany; two examples were used to shuttle VIPs and serve as support cars in various long-distance events like the marathon. With a front engine, rear-wheel drive layout, the 1602 Elektro-Antrieb used one 32 kW (43 hp) motor, had a claimed top speed of 100 kmh (62 mph) and acceleration of 0–30 mph (0–48 kmh) in 8.0 s. It never made commercial production.

Following the 1970s, the Middle East wars and hike in gasolene prices, the impetus to develop a viable electric car moved to the price and availability of gasoline.

Launched in 1974, the 'CitiCar' was as close to commercially viable as any achieved. Inspired by its Club Car's golf cart design, Sebring-Vanguard produced its first electric vehicle, the Vanguard Coupe (sometimes referred to as the EV Coupe), in 1974. This evolved into the 'CitiCar' after the EV Coupe did not find success. The CitiCar was a small wedge-shaped basic short range electric vehicle. But by 1976, enough CitiCars were produced to promote Sebring-Vanguard to the position of being the USA #6 auto manufacturer after GM, Ford, Chrysler, AMC, and Checker Motors Corporation. Production of the CitiCar continued until 1977 with about 2300 CitiCars produced. Commuter Vehicles, Inc. purchased the CitiCar design, and renamed the vehicle Comuta-Car. Production of this upgraded version began in 1979 and Commuter Vehicles, Inc. produced an estimated 2144 Comuta-Cars and Vans.

In 1974–76, Zagato, maker of marques such as Alpha-Romeo, manufactured a rear engine microcar called the 'Zele' (known in the USA as the 'Elcar'). The Zele employed a 4 position speed selector and a 2 position foot pedal providing six forward speeds, two reverse speeds, and a range of approximately 50 miles (80 km). The Zele 2000 also featured a boost switch which, once at top speed, weakens the motor's magnetic fields in the field coils to produce less torque but a greater top speed. Its top speed was between 25 and 30 mph. The American importer, Elcar Corporation, also created an elongated 4-seat prototype called the Elcar Wagonette. In total about 500 were made and sold.

In 1974, came the Jeep DJ-5E 'Electruck', a delivery vehicle produced mainly for the US Postal Service, able to operate with the limited range required for urban postal collection delivery, this vehicle can be considered one of very few successes of its time. Based on the same 81-inch wheelbase, two-wheel drive chassis and body shell as the familiar Jeep postal vehicle, the Electruck was essentially identical in exterior appearance. Attracted to the potentially lower operating costs, the US Postal Service purchased 342 Electrucks, and reportedly, they gave more or less acceptable service—within their limitations—until 1983 when they were retired. By then, the postal authorities had determined that vehicle service life was the important factor in managing costs, and commissioned the Grumman LLV. The LLV stood for Long Life Vehicle.

In 1977, Sweden's Volvo experimented with an electric prototype, with the express aim to reduce city pollution. Simply known as the Volvo Electric Car, it had two seats, a top speed of 70 kph/43 mph and a driving range of about 2 h (140 km/87 miles), but took 10 h to recharge. Only two concepts were built.

In 1979 GM tried again, this time with the 'Electrovette'. Work on the Chevrolet Electrovette Concept began in 1976, and it was released to the public in 1978. The Electrovette was designed primarily to be cheap to own and operate. Based on the production Chevrolet Chevette, the Electrovette featured a shortened wheelbase, and a battery pack positioned where the rear seat would normally be located. A 50 kW (63 horsepower) motor was shoehorned into the space normally occupied by a transmission. It was designed to be powered by nickel-zinc (NiZn) batteries, but performance was disappointing, and they were replaced by less-expensive lead-acid batteries. Electrovette was claimed to have a range of 50 miles at a constant 30 mph, and had a top speed of 53 mph. It made it to AGMs, motor shows, and press releases, but no further. At the time, the Electrovette was retired, and GM engineers stated that a battery breakthrough of some sort would be required before an affordable, consumer-friendly electric vehicle could be built in volume.

In 1980, the Soviet Union, showcasing at its hosting of the Olympic Games, introduced the VAZ 2801, an electric car based on its 1978 lead-acid battery-powered VAZ 2101 'Lada'-based prototype but using lighter nickel-zinc batteries and an aluminium frame, it had a top speed of 54 mph and a range of 68 miles. However, impracticalities such short range in such a huge country, the lack of a recharging network, and cost, meant that it failed to find success.

In 1985, serial UK inventor Clive Sinclair launched one of the more eccentric contributions, the 'Sinclair C5'. Half car, half pedal cycle, the C5 emerged from an earlier project to produce a small electric car called the C1. After a change in the law, prompted by lobbying from bicycle manufacturers, Sinclair developed the C5 as an electrically powered tricycle with a polypropylene body and a chassis designed by Lotus Cars. It was intended to be the first in a series of increasingly ambitious electric vehicles, but the development of the follow-up C10 and C15 models never progressed further than the drawing board, mostly due to the poor public response to the C5.

On 10 January 1985, the C5 was received less than enthusiastically by the British media. The vehicle's limitations—a short range, a maximum speed of only 15 miles per hour (24 kmh), a battery that ran down quickly and a lack of weatherproofing—made it impractical for most people's needs. Out of 14,000 C5s made, only 5000 were sold before its manufacturer, Sinclair Vehicles, went into receivership.

During the 1990s, BMW showcased its E1 electric vehicle to demonstrate its capabilities with electric motors. Unfortunately, the car caught fire a few times during various shows, which resulted in it not receiving the attention it deserved. As a result, the E1 never saw mainstream production. Despite this setback, the E1's specifications are still impressive for a car produced 30 years ago. It had a range of 150 miles, which was impressive for its time.

The 1990 saw a spurt of new electric vehicles for an entirely new reason—the so-called CARB 2% Act. Spurred by the need to reduce rising air pollution and several smog incidents occurring in its cities, in September 1990, the California Air Resources Board adopted a pioneering mandate that would require 2% of new cars sold in California beginning with 1998 models to be exhaust-free—essentially a requirement for electric cars. To be more precise they mandated that each of the seven largest automakers in the USA, have to ensure that 2% of their fleet would be emission-free by 1998, increasing to 5% by 2001 and 10% by 2003, based on consumer demand. Automobile manufacturers were appalled and sceptical, but had to show willing and launch EV models as an option to buy time while they opposed the ruling through the courts. GM was particularly appalled because it was widely believed that CARBs confidence to make this mandate was based in no small part on claims GM was making about the intended scale of production for its new 'Impact' EV. (see below).

Chrysler had a special interest in producing minivans, particularly electrified ones. In 1993, it launched the Chrysler 'TEVan', which was powered by nickel-iron or nickel-cadmium batteries. Its performance and technology were not revolutionary. The TEVan had a maximum speed of 70 mph and a range of 50 miles. Sales were a challenge due to its high price tag of $120,000. Only 56 TEVans were ever made.

Nissan has been researching and developing electric vehicles since 1947, when the Tokyo Electro Automobile Co. produced the Tama electric vehicle (see above). In 1991, Nissan began selling the Cedric EV, and then the Avenir EV in 1994, primarily to companies. Nissan decided to adopt the world's first commercialised lithium-ion batteries by Sony Corporation and started joint development with Sony in 1992. Launched for a limited lease programme in 1997 as the 'Prairie Joy EV', the world's first EV on the market with lithium-ion battery, it showed sufficient performance with the max. Speed of 120 kmh and a claimed range of over 200 km. But the Prairie Joy was not further launched commercially.

In 1998, Nissan released the 'Altra EV' which was based on a minivan R'nessa. The Altra had a 62 kW electric motor and run on lithium-ion batteries (manufactured by Sony) and claimed a range of 230 km (140 miles) with a time of 5 h to recharge, and over 1000 charging cycles.

This Altra was equipped with several advanced features, such as a flat floor and underfloor batteries, non-contact charging system, air conditioner with remote control, EV-dedicated digital meters, lithium-ion battery, and permanent magnet synchronous motor. All of these technologies were later implemented in the first-generation Nissan Leaf in the twenty-first century. But the Altra was not launched commercially.

The 'Chevrolet S-10' EV was an electric vehicle that was introduced at a similar time. The Chevrolet S-10 Electric was introduced in 1997, becoming the world's first electric pickup truck from the original manufacturer. It was updated in 1998, and then discontinued. The Electric S-10 EV was equipped with an 85-kw (114 hp) three-phase, liquid-cooled AC induction motor, based on GM's EV1 electric coupe (see below). The EV1 had a 100 kW motor; but GM reduced the S-10 EV's motor output because of the additional weight and drag of the truck so as not to overstress the batteries. It was marketed primarily to utility fleet customers. However, the S-10 EV was not revolutionary. The car was quite heavy, weighing a combined 1400 pounds due to its heavy batteries. Although it had a respectable 114 hp., the vehicle's range of 45–60 miles was clearly inadequate.

Citroen developed the 'AX électrique' the early nineties, with some prototypes on the road in 1993. It formally launched in November 1995 proffering a (peak power) 20 Kw engine (nominal power 11 Kw) with an urban range of about 75 km/47 miles and a top speed of 91 kmh/56 mph which it could reach in a little over 8 s. niCad battery, with a proprietary fast-charging system. A total of 374 were produced between 2003 and 2006 (mostly to utility companies) when it was replaced by the 'Saxo électrique' which sold 2174 units between 1996 and 2003, and the 1997 'Berlingo électrique' with 1749 vehicles built.

The Saxo électrique had the same drive system as the AX électrique, but also a 2.5 gallon petrol tank to power the heater. You could buy the car (about US$ 21,500) but you had to rent the batteries from the manufacturer for about $100 per month. It had no spare tyre because there was not any room for it. There was a Peugeot badged equivalent model.

The Citroën Berlingo électrique is a battery-powered version of the first-generation Berlingo range of vans, built and sold between 1998 and 2005. Featuring a 162 V Saft NiCd battery, a 28 kW Leroy Somer electric motor, it claimed a maximum speed of 95 kmh (59 mph), with a maximum range of 95 km (59 miles) in typical driving. 1749 vehicles were built, largely in the early 2000s.

Honda had been experimenting with electric vehicles since the 1980s. However, the Honda EV Plus (ZA1) was built specifically to meet California's Air Resources Board (CARB) requirement. This EV car used a nickel-metal battery (NiMH) and could cover a distance of 60–80 miles on a single charge. It boasted an impressive charging time, with 80% of the recharge being achieved in just 2 h. About 340 EV Plus models were produced and released. Production of the EV Plus was discontinued in 1999 after Honda announced the release of its first hybrid electric vehicle, the Honda Insight. The NiHM battery version was priced at $ 54,000 which made it quite expensive, and the batteries needed to be replaced every 3 years. At the time, Honda stated that they had no plans to create another EV until the technology advanced enough to reduce charging times.

The Ford Ranger EV was an electric vehicle produced by Ford Motor Company from 1998 to 2002. It was built using a light truck chassis from the Ford Explorer. A battery electric compact pickup truck which was the automaker's first all-electric production vehicle. It was produced starting in the 1998 model year through 2002 when it was discontinued. Most vehicles were sold with 23 or 26 kWh nickel-metal hydride batteries (NiMH). A few vehicles with lead-acid batteries were sold. Most units were leased for fleet use (mainly in California), and leases were terminated in 2003–04, and the vehicles were recalled providing a range of up to 65 miles on a steady speed of 65 mph or up to 120 miles driven carefully below 45 mph. The cost of this vehicle was $50,000. Early in the vehicle's life cycle, the NiMH Ranger EV had some issues that were later addressed by Ford. However, there were some problems with the range achievable towards the end of the 25,000-mile service life of the batteries, and Ford chose not to fix this problem.

The first-generation Toyota RAV4 EV, launched in 1997, was a revolutionary car designed to demonstrate that Toyota has taken CARB seriously and to showcase the potential of electric vehicles. The first generation was leased from 1997 to 2003. A total of 1484 were leased and/or sold in California to meet the state's mandate for zero-emissions vehicle. A small number were also sold or leased in fleet sales in other states. It claimed a range of 124 miles and featured nickel-metal batteries that weighed 450 kg, which had a lifespan three times longer than lead-acid batteries. Moreover, these batteries could be recharged in just 10 h (a second-generation Toyota RAV4 EV was launched in 2012 and sales continued until it was discontinued in 2014. But more about that later on).

Finally in the 1990s, we turn our attention back to GM (remember from above the 1960s Electrovair I & II and the 1970s Electrovette). At the same time as fighting the CARB 2% requirement in the courts, GM made a determined effort to commercialise the electric vehicle.

The team at General Motors (GM) created the EV1 electric car. The (somewhat unfortunately named) 'Impact' prototype was unveiled at the LA Auto Show in January 1990. A highly modified variant set the speed record for an EV in 1994. The General Motors EV1 was a battery electric car developed from the 'Impact' and produced from 1996 until its demise in 1999.

A subcompact car, the EV1 was not an adaptation of another existing model, but was designed 'no expense spared' from the bottom up as a purpose-built EV, designed for the introduction of mass produced and purpose-built battery electric vehicles. It was based on the 'Impact' prototype. The car was powered by 32 lead-acid rechargeable batteries.

In April 1990, GM Chairman, Smith announced that the Impact would become a production vehicle with a goal of 25,000 vehicles.

In 1994, GM initiated 'PrEView', a programme in which 50 handcrafted Impact electric cars would be loaned to drivers for durations of 1 to 2 weeks, with the stipulation that their feedback and experiences would be documented. Volunteers were required to possess a garage suitable for the installation of a high-current charging unit by an electric company. Driver response to the cars was favourable, as were reviews by the automotive press. According to Motor Trend, the Impact 'is precisely one of those occasions where GM proves beyond any doubt that it knows how to build fantastic automobiles. This is the world's only electric vehicle that drives like a real car'. (source: Wikipedia). Despite the good reception, as highlighted in a front-page feature in The New York Times, GM appeared to be less than enthusiastic about the prospect of having created a thriving electric car:

In a front-page article, the New York Times claimed that 'General Motors is preparing to put its electric vehicle act on the road, and planning for a flop. .... the face of a California law that requires that [two] percent of new cars be "zero emission" vehicles beginning in 1997, it has done its best but that the vehicle has come up short. ... Now it hopes that lawmakers and regulators will agree with it and postpone or scrap the deadline'.

Dennis Minano, GM's vice president for Energy and Environment, questioned whether consumers desired electric vehicles. Robert James Eaton, chairman of Chrysler, also doubted the readiness of mass produced for electric cars, stating in 1994 that 'if the law is there, we'll meet it ... at this point of time, nobody can forecast that we can make an electric car'. (source: Wikipedia).

Following the PrEView initiative, work on the GM electric car programme continued until the design had evolved into the GM EV1 in 1996. The first-generation EV1 had lead-acid batteries and claimed a range of 70–100 miles. 660 vehicles were produced and offered for lease only, explicitly prohibiting the option to buy under a contractual provision (with a suggested retail price listed at $ 34,000).

Similar to the PrEView programme, lessees were pre-screened by GM, with only residents of Southern California and Arizona initially eligible for participation. Leasing rates for the EV1 ranged from $ 399 to $ 549 a month. The car's debut was marked by a significant media event, featuring an US$8 million promotional campaign incorporating prime-time TV commercials, billboards, a dedicated website, and an appearance at the premiere of the Sylvester Stallone film Daylight. Among the initial lessees were notable figures such as celebrities, executives, and politicians. At the release event, 40 EV1 leases were signed. Deliveries began on December 5, 1996. In the first year on the market, GM leased just 288 cars (source: Wikipedia).

In its initial stages of production, most EV1s were leased to consumers in California, Arizona, and Georgia. Despite favourable customer reception, GM found that electric cars occupied an unprofitable niche of the automobile market and in 2001 GM terminated the EV1 programme, and all current leases and crushed most of the cars, disregarding protests from customers many of who wanted to buy out their leases.

Since its demise, the EV1's cancellation has remained a subject of dispute and controversy.

Summary 1950–2000

To put this period into perspective, in the 1950s just under 100 million cars were made. In the 1960s, a little under 170 million cars were made. By the 1980s, this had grown to over 300 million and by the 1990s to over 370 million. In the total period 1950–2000, this adds up to over 1.25 billion vehicles (apx 1,255,000,000).

The pages above summarise the considerable efforts of existing car manufacturers, inventers, entrepreneurs, research institutes, etc. to produce a commercially viable BEV. There were good reasons and motivation to do so: availability of gasolene, price of gasolene, city pollution levels, legal requirements for zero tailpipe emissions vehicles.

Excluding milk carts and golf carts, summarising the sales described above, we would be optimistic if we summarised more than 5000 BEVs sold or leased in half a century (0.0004%) and all but two models (the RAV4 EV and Citroen Saxo électrique) discontinued by 2002, and the RAV4 EV, which achieved 1484 unit leases, and Saxo électrique, 2174 units, and the only two models that could claim any limited commercial viability, were both discontinued in 2003.

I think it is fair to conclude that for all the effort and investment made, in the period 1950–2000, the BEV was not fit for purpose.

The major problems not overcome by all were range, recharging time, battery life, and cost (although other design issues were also encountered by many).

By 2000, the laudable 2% ambitions of CARB had no way they could be met.

GM, and especially the truncation and demise of its AV1 project, took a lot of flak for 'killing off the BEV'. But GM invested more than most in this period to make the BEV a commercially viable concept. But a commercial company cannot escape the facts.

The concept of the battery electric vehicle failed in 1950–2000 for the same reasons as its demise and fall in the first half of the century. Despite new pressing needs, the technical solutions were just not up to the job. Better solutions were available.

It would take something to change fundamentally to alter that conclusion.

A New Millennium

In the 1960s and early 1970s, people in general were more worried about when the next ice age would come than they were about global warming. In the 1960s, scientists were becoming increasingly aware that estimates of global temperatures showed cooling since 1945. Speculation ranged from increased cloud cover due to pollution to the cooling effect of aerosols to natural climate cycles. But by the time the idea of global cooling reached the popular public press, temperatures had stopped falling, and fears of the opposite were emerging.

1970s Environmental Movement: The first Earth Day in 1970 and subsequent environmental regulations raised public awareness about various environmental issues, including climate change and raised awareness of the work of Jean-Baptiste Fourier, John Tyndall, Eunice Newton Foote, Svante Arrhenius, Guy Stewart Callendar, Charles David Keeling, and others to a sceptical public. At the follow-up 1979 World Climate Conference, scientists called for international action to address climate change. The conference (full of grant hungry scientists) emphasised the need for continued research and global cooperation.

In 1988, NASA scientist, James Hansen, testified before the US Congress, stating with high confidence that human activities were causing global warming. This was a pivotal moment that brought climate change into the political arena.

Growing awareness and pressure led to the formation of the 'Intergovernmental Panel on Climate Change (IPCC)' in 1988. Established by the UN, the IPCC provided comprehensive assessments of climate science. Its reports have been crucial in informing global policy and negotiations and provided further negotiations which led to the 'Kyoto Protocol (1997)': The first major international treaty to reduce greenhouse gas emissions. It set binding emission reduction targets for developed countries.

The Kyoto Protocol is an international treaty that commits its parties to reduce greenhouse gas (GHG) emissions, based on the scientific consensus that global warming is occurring and that human-made CO_2 emissions are driving it. It was adopted on December 11, 1997, in Kyoto, Japan, and entered into force on February 16, 2005. While there is little doubt that global warming is influenced by human activity, even though the nominal effect of each type of activity is arguable, CO_2 emissions are recognised as the main culprit.

So when the new millennium dawned, and business got over worrying about 'the Millennium Bug' crashing their computer systems, they were faced with the problem with how to reduce greenhouse gas emissions, and quickly.

The first response to the Kyoto Protocol is a cautionary tale.

If BEVs were not practical yet, Governments encouraged a switch from petrol to diesel. Diesel was seen as a way of driving cost-efficiently and contributing to save the planet. Government, industry, and science united together to persuade us that cars running on diesel would help us cut our CO_2 emissions and contribute to achieving the Kyoto targets. Advances in diesel engine technology caused expansion of its appeal, particularly in Europe. Drivers were attracted to the fuel economy of diesel engines and were encouraged to switch by Governments, in order to help achieve their Kyoto targets. Diesel cars were thought to emit less CO_2 than their petrol equivalent. Many governments cut fuel tax on diesel to further encourage the changeover.

It wasn't until 2012 that the truth came to light. It turns out that the nitrogen oxides and dioxides (NOx) and particulate matter emitted by diesel exhausts were silent killers. The World Health Organization declared diesel exhaust to be a carcinogenic, a cause of lung cancer in the same category as asbestos and mustard gas. And to make matters worse, auto manufacturers had been cheating with their emissions testing. As David King, the UK government's former chief scientific adviser on climate change, admitted in 2017, ministers had made a huge mistake by promoting diesel. They had trusted the car industry when it said the fuel was clean. 'It turns out we were wrong', he said.

A more successful outcome was the appearance of the hybrid car. Coincidentally in 1997, the Toyota Prius was launched. Hybrid powertrains switch from one power source to another to maximise both fuel efficiency and energy efficiency. An electric motor is more efficient at producing torque, or turning power, while the combustion engine is better for maintaining high speed. Combined with regenerative braking, it results in improved efficiency, lower emissions, and therefore reduced running costs.

The Prius was more expensive to buy than its petrol competitors, but its economy made it competitive in the longer run, and it was particularly advantageous for high mileages. Owners too could feel that they were doing something for the planet. The Prius was a hit from day one, and, 27 years later, much upgraded, is still a sales-leading hybrid today. It remains dearer than its petrol only equivalent, but is less expensive than its BEV equivalent.

The Prius was soon followed by others. In Europe, Audi launched the first PHEV—'Duo III' (but it was priced too high to sell many units). In 1999, Honda launched its 'Insight' model which it replaced with the 'Honda Civic Hybrid' in 2022. Lexus launched the first hybrid SUV, the RX400, at the end of 2002. Ford entered this arena in 2004 with the 'Ford Escape Hybrid' using licenced Toyota technology. GM entered with the 2007 model 'Saturn Vue Green Line' and by 2010 hybrid was becoming either the principal powertrain or an option for many models around the world, for all but bottom of the market models. But, I digress, this work is about BEVs.

Very little new seemed to be happening on the BEV front except the cessation of models reported above.

Volvo, a Swedish company with a reputation for making safe vehicles, in 2001, developed the 'Volvo SCC' (Safety Concept Car). Its features included forward collision warning, blind spot monitor and alert, lane departure warning, adaptive cruise control, head-up display, flashing brake lights, and steering linked headlights. Common features in today's cars, but revolutionary in 2001. The SCC also featured communication with the car via a mobile phone. The SCC was first unveiled at the 2001 Detroit Auto Show, and it led to development of the C30 which began in 2003. But it wasn't until the end of the decade that an electric version was produced. This led in the early 2010s to a smaller series of fully electric Volvo C30s which were developed for company car drivers and in the government. The cars were charged from a wall socket, and the range was about 150 km. No commercial product, however, emerged before 2010.

But Tesla was incorporated in July 2003 by Martin Eberhard and Marc Tarpenning as Tesla Motors. In February 2004, Elon Musk joined as the company's largest shareholder; in 2008, he became chief executive officer. In 2008, the company began production of its first electric car model, the Roadster sports car, the Tesla Roadster was a battery electric sports car, based on the Lotus Elise chassis, produced by Tesla Motors from 2008 to 2012. It was powered by lithium-ion battery cells, and the first production all-electric car to travel more than 244 miles (393 km) per charge. It had lightening acceleration, claiming to accelerate from 0 to 60 mph (0–97 kmh) in 3.7 or 3.9 s depending on the model; and a top speed of 125 mph (201 kmh). Tesla claimed that the Roadster's efficiency was 120 miles per gallon gasoline equivalent (28 kWh/100 miles) (2.0 L/100 km). It was, in short, the first BEV to come close to the performance requirements of the average user. But at around US$ 110,000 was expensive and came nowhere near what the average user was prepared to, or able to, pay; and required a home charging station in addition.

Before it officially commenced BEV production, and following on from the Prairie Joy EV and the Altra EV in the 1990s, Nissan developed three prototype battery electric vehicles: the EV-01 and EV-02 both based on the Nissan Cube, and the EV11, based on the Nissan Tilda. These evolved over the decade and resulted in the Nissan 'Leaf'. The Nissan Leaf did not begin production until October 2010 (at the Oppama facility in Yokosuka) and went on to be a commercial success. Production began in the UK in March 2013. The Leaf was powered by a large battery pack composed of 192 lithium-ion cells developed in collaboration with

NEC. The electric motor produced 80 kW (110 hp) with a range of about 100 miles (160 km). Recharging needed 16 h on 110 V or 8 h on 220 V power. Fast charging was also available with a purpose designed charger, which could restore 80% of the battery capacity in approximately 30 min. The launch price of the Leaf in 2010 was apx US$ 33,000 (but by this stage the US government was providing a grant of US$ 7500 to incentivise BEV purchases).

So while the overt focus of the 2000s was less polluting petrol vehicles, and encouraging diesel engines, there were very few new BEVs released. Indeed, when it became clear that battery electric vehicles were not meeting necessary cost and performance goals, the CARB Zero-Emissions Vehicle programme was modified to provide credit for new types of clean conventional vehicles and for vehicles that employed advanced technology, but which did not have zero-emissions. Hence the rapid development of hybrids. But, behind the scene, the seeds of change were being sown, and sown not the least in China.

China was by this time the fastest growing auto market in the world. In January 2003 (the same year as Tesla was founded), the founder of BYD (a battery manufacturer) Wang Chuanfu founded BYD Auto after acquiring and renaming a small automotive manufacturing company. Chuanfu acquired Qinchuan with the intention of developing battery-powered electric vehicles, using BYD's expertise in battery manufacturing and creating sales expansion opportunity also for its sister company. Meantime, BYD Auto expanded by making Japanese and western lookalikes for the rapidly expanding Chinese market.

At the 2006 Beijing Auto Show, BYD launched the battery electric version of its F3 model, the BYD F3e. The car had a BEV range of more than 300 km (186 miles) and was planned to be produced within 3 years from 2007. But in December 2010, BYD Auto deputy general manager Wang Jianjun confirmed that the company had cancelled plans for its production due to the lack of support in charging infrastructure. But according to BYDs founder, electric vehicles would be BYD's stepping stone to enter foreign markets such as Europe and North America.

In 2008, Wang Chuanfu set a goal for the company to lead the Chinese automotive market by 2015 and to be the largest car manufacturer in the world by 2025.

In January 2009, BYD introduced its first production BEV, the e6 at the 2009 North American International Auto Show in Detroit. It was powered by a battery pack codenamed 'Fe', which provides a claimed electric range of 400 km (249 miles) While plans to sell the vehicle to the general public in the USA were shelved, the company sold the e6 in the USA in a limited number as a fleet vehicle.

(Source: Wikipedia)

In 2010, BYD Auto Industry Co. Ltd. and Daimler AG, now known as Mercedes-Benz Group AG, formed a 50–50 joint venture named 'Shenzhen BYD Daimler New Technology' with a brand named 'Denza' to focus on research and development of new energy vehicles. The brand showcased the Denza EV concept model at Auto China in April 2012. (Denza was later restructured in 2021, when BYD took control of the brand by taking a 90% stake in the venture.)

(Source: Wikipedia)

The Chinese economy/economic strategy is led by 5-year plans called the 86/3 Program, which provides financial incentives to achieve its stated goals. The tenth revision of the 86/3 Program was issued in 2001 and included the development of BEVs, hybrids, and fuel cell vehicles. The National Development and Reform Commission subsequently published the Auto Industry Development Policy in 2004. Later in 2004, 16 Chinese state-owned companies formed an electric vehicle industry association in Beijing called CEVA. The goal of the association was to integrate technological standards and create a mechanism through which stakeholders share information in order to develop upmarket BEVs. The companies were anticipated to invest a total of US$14.7 billion on the growing electric vehicles market by 2012.

By 2008, sales volume of Chinese BEVs reached 366 units in the first half of 2008, and Chinese automakers then provided around 500 independently developed new energy and fuel efficiency vehicles to serve the Beijing Olympics (although how many of these were BEVs is unknown).

In 2009, 13 cities (Beijing, Shanghai, Chongqing, Changchun, Dalian, Hangzhou, Jinan, Wuhan, Shenzhen, Hefei, Changsha, Kunming, and Nanchang) were shortlisted as pilot cities for BEV usage and the State Council approved the 'Auto Industry Restructuring and Revitalization Plan' to invest RMB 10 billion (US$ 1.50 billion) for new e-vehicle industrialisation and RMB 20 billion (US$ 3 billion) in automotive technology development to improve air quality and reduce reliance on fossil fuels (China has set itself a target of carbon emission neutrality by 2060). The government then announced a 2-year pilot programme of subsidising purchasers of alternative energy cars in five cities (Shanghai, Changchun, Shenzhen, Hangzhou, and Hefei) of RMB 60,000 for battery electric vehicles (a little over US$ 8000) and RMB 50,000 for plug-in hybrid vehicles (PHEV).

The Growth and Influence of the Chinese Market

In 1990, there were just 5.5 million cars in China (by comparison, in 1990, in the comparatively tiny UK, there were over 21 million cars). By 2024, the number of cars in China is approaching 300 million cars. China is now the world's largest car park, and by far the largest market for new car sales (even before considering its capacity to export cars). If China reaches western countries car population density, its car park will grow to over 400 million cars.

By 2011, there were several major electric vehicle R&D and manufacturing companies active in Chinese market including:

BYD

Continuing the summary of BYD (see above) from 2010 to date. In August 2013, BYD launched the 'Dynasty' model range—a plug-in hybrid electric variant of the petrol-powered BYD Surui, which it named the BYD 'Qin' and went on to become the best-selling plug-in electric vehicle of 2014.

The first-generation BYD 'Tang', the second vehicle in the 'Dynasty' range, was introduced at the 2014 Beijing Auto Show and was available as a plug-in hybrid only. The BYD Tang was a mid-size crossover SUV. The Tang was the top-selling plug-in electric passenger car in China in 2016.

The second-generation BYD Tang was launched at Auto China in April 2018, and offered an 82.8 kWh battery electric variant. That variant went on sale in Europe (in Norway) in 2021.

According to Wikipedia, during the period before 2020, BYD relied on government subsidies to create profit in selling plug-in hybrid and battery electric vehicles. For instance, throughout 2016, the company recouped about US$ 1 billion in new energy vehicle subsidies for many of the nearly 100,000 new energy vehicles it sold. But note the volume sold in 2016. So far in this work 'success' has been measured as a few thousand units spread over several years. By 2016, BYD were selling really big sales volumes.

But the subsidies BYD received were also equivalent to a little over a fifth of its US$ 5 billion in revenues from new energy vehicle sales that year. So when the government reduced subsidies between 2017 and 2019, BYD was negatively impacted, experiencing a sales slowdown in that period with net profit falling sharply for three consecutive years.

The launch of the BYD Atto 3/Yuan Plus in 2021 helped increase the sales of BYDs globally; and it is currently one of the best-selling electric cars in Australia. The Atto 3/Yuan Plus is an SUV that uses a 150 kW front-wheel drive electric motor powered by BYD's proprietary lithium iron phosphate (LFP) 'Blade' battery. It offers two battery pack options—a 49.92 kWh unit capable of travelling 345 km (214 miles), while the 60.48 kWh battery pack is rated at 420 km (261 miles). The vehicle is built on 400-volt electrical architecture and supports maximum AC charging (Type 2) of 7 kW (standard) or 11 kW (option) as well as DC fast charging (CCS2) at a maximum rate of either 70 kW (Standard Range) or 80 kW (Extended Range). At 80 kW, charging from 0% to 80% can be completed in 45 min.

BYD launched the 'BYD Han' large sedan, powered by BYD's 'Blade' battery, in early 2020 with sales commencing in the second half of that year, with an option of plug-in hybrid electric (Han DM) and battery electric (Han EV) variants. The range-topping variant of the Han EV was claimed to be the fastest electric car in China, while the DM version was the fastest hybrid sedan.

In April 2021, BYD introduced its 'e-Platform 3.0', a third-generation platform for BEVs that integrated and standardised core components along with a new body structure, new electrical architecture, and operating system. The platform began production starting with the BYD Dolphin and the BYD Yuan Plus that were announced in August 2021, also establishing the 'Ocean' Series line of products.

BYD experienced a substantial surge in vehicle sales from 2020 to 2023. The company sold 3,024,417 vehicles globally in 2023, a seven-fold increase compared to the 2020 figure of 427,302 vehicles. (source: Wikipedia).

In the first half of 2022, BYD surpassed Tesla as the world's largest plug-in electric vehicle manufacturer by selling 641,000 vehicles. In September 2022, BYD became the first carmaker in China to build 1 million new energy vehicles in a single year.

By 2023, BYD continued to be the world's largest PHEV manufacturer and the second largest BEV manufacturer (after Tesla), with global market shares of 21.4% and 15%, respectively. (source: Wikipedia).

BYD announced its entry to Europe in May 2020, starting with Norway. The first batch of 100 BYD Tang EVs equipped with the blade battery was sent to Norway in June 2021.

In January 2023, BYD established its second premium brand called Yangwang by introducing the Yangwang U8 plug-in hybrid electric large SUV, and the Yangwang U9 battery electric supercar.

In April 2023, BYD released its smallest and cheapest BEV called the Seagull. It features a 4-seater, 5-door configuration in a hatchback body style, while being priced in a segment dominated by smaller 3-door cars. In China, there are three variants offered: the 'Vitality Edition', a higher specification 'Freedom Edition' and a top-specification variant called the 'Flying Edition'. All three models are powered by a single 55 kW (74 hp) motor with a top speed of 130 kmh (81 mph) and with a 0–100 kmh of 13 s.

Both the Vitality and Freedom models are built with a 30 kWh lithium-ion phosphate blade battery, enabling a cruise range of 305 km (190 miles); the Flying model sports a larger 38 kWh battery and a range of approximately 405 km (252 miles). The battery can be recharged via a household AC power plug at 6.6 kW, or via a DC fast charger at 30 or 40 kW (depending on the model), during which it reportedly can be charged from 30% to 80% capacity in 30 min. (source: Wikipedia).

On 29 November 2023, BYD produced the 200,000th Seagull after only seven months in the market. Due to its low cost of production and its low selling price at below RMB 90,000 or US$ 12,000, the Seagull potentially offers to compete at prices below many of its combustion engine equivalents.

SAIC

Originating from an automobile assembly factory established in Shanghai sometime around World War II, SAIC, was a small company in the 1970s, growing with the domestic automotive market through a cooperative agreement made with Volkswagen in 1984 followed by the formal establishment of Shanghai Volkswagen Automotive Co. Ltd. in March 1985 and the company quickly established itself as one of the leading Chinese automakers.

SAIC began exploring electric mobility in the early 2000s, as with others, under the 20th revision of the 86/3 Program (see above), but the company's early efforts focused on hybrid vehicles and the development of basic electric vehicle (EV) technologies. SAIC formed key partnerships with international automakers to enhance its EV capabilities:

- SAIC-GM: In 1997, SAIC entered a joint venture with General Motors (GM). This partnership later played a crucial role in the development of electric and hybrid vehicles.
- SAIC Volkswagen: Another significant joint venture was established with Volkswagen in 1984, which contributed to technology exchange and innovation.
- In 2006, SAIC acquired the British brand Rover, which led to the creation of the Roewe brand, and by doing so, also acquired the iconic British brand MG, which has subsequently been used to market its electric vehicles globally.

SAIC's first mass-produced electric vehicle was the Roewe E50, a 4-seater supermini sized hatchback launched in 2012 and sold from 2013. The E50 had a 47 kW motor and an 18 kWh battery giving it a range of about 120 km (75 miles) and a top speed of up to 120 kmh (75 mph) and was designed for urban commuting in China. Pricing started at RMB 234,900 (US$ 37,589) before available government incentives, and SAIC's goal was to sell 1000 units within the first year in the market. The E50 provided the basis of MG Dynamo EV, a concept car shown in the UK during spring 2014. The Dynamo had a 71 bhp electric motor developing 155 Nm of torque, and a range of 50 miles.

Launched at the 2018 Guangzhou Auto Show, and sold from 2019, the MG EZS/ZS EV, a BEV version of its popular MG ZS model, marked a significant milestone for SAIC's international EV strategy. It offered a 44.5 kWh battery, and a front positioned motor producing 141 hp. (105 kW) to the front wheels. It was launched in the UK, parts of Europe, Thailand, China, India, New Zealand, Australia, and Pakistan.

It was revealed in Australia in November 2020 as the cheapest electric car in the market of all time. MG Australia predicted a sales target of 3000 units for the 2021 model year.

In fall 2021, a facelift version also equipped with a larger, 72.6-kWh battery was launched in Europe. This compact electric SUV, launched in various markets including Europe, offered a competitive range, affordability, and modern features, making it popular among consumers. With its MG badge, it gained particular success in Europe. SAIC has continued to expand its EV lineup with models such as the Roewe Marvel R, the MG E4, the MG5 Long Range (and a plug-in hybrid).

In 2020, MG sold approximately 23,000 BEVs globally. MG saw a significant increase in BEV sales in 2021, selling around 42,000 units, 67,000 units in 2022, and around 85,000 units in 2023. (source: Chat GPT).

These figures reflect MG's growing presence and success in the global BEV market, supported by its strategic focus on expanding its electric vehicle lineup and improving its technology offerings. The company continues to invest heavily in research and development, focusing on improving battery technology, increasing vehicle range, and enhancing automated (ADAS) driving capabilities. The company is also investing in building a robust charging infrastructure to support its growing EV fleet.

Changan Automobile

Changan Automobile's history is marked by a transformation from a military enterprise to a leading player in the global automotive market. Through strategic partnerships, technological innovation, and a strong focus on electric and smart vehicle development, Changan continues to evolve and adapt to the changing landscape of the automotive industry. Currently, it sells around 2.25–2.5 million vehicles per year. Changan Automobile has been gradually increasing its focus on BEVs over the years. In 2015, it launched the Changan CS15EV is a BEV based on the Changan CS15 subcompact SUV; powered by an electric motor producing 75 hp. and 170 nm of torque, setting a 110 kmh top speed and a range of 300 km. According to the official website, the prices of the Changan CS15EV ranged from RMB 190,000–195,000 and sold around 5000 BEVs.

In 2016, Changan sold about 8000 BEVs launching the ChangAn BenBen Hatchback, which is still on sale). The popular Changan BenBen EV followed the Eado EV in October 2016 with the Eulove MPV van and Eado PHEV in 2017. Various international automakers including PSA, Ford, and Suzuki have formed joint ventures with the company.

Changan's EV strategy is underpinned by its '6321' new energy technology goals, which includes a pure electric vehicle range of 600 km, the technology to be electric, intelligent, and lightweight, consisting of two pure electric vehicle platforms. Changan and Baidu entered into a cooperation agreement in 2017 to develop autonomous technology. In 2017, it sold around 12,000 BEVs in 2018 18,000.

In 2019, it sold about 25,000 BEVs launching the 'Changan CS15 E-Pro' at the 2019 World Intelligent Network Automotive Conference and the seventh China International New Energy and Intelligent Networking Automotive Exhibition (IEEV China 2019). It is essentially the Changan CS15EV with performance improvements included. The CS15 E-Pro has a 120 kW (160 hp) motor and 48.3 kWh battery.

This generated sales of about 45,000 in 2020 which saw the launch of the BenBen 'E-Star', an economy model selling at just over RMB 100,000 (US$ 14,000); rising to 2021, 50,000 (driven by the introduction of popular models and expanded charging infrastructure); 2022, 70,000; rising to an estimated 100,000 vehicles in 2023, driven by strong growth momentum in the BEV segment, supported by technological advancements and strategic partnerships. The comprehensive Changan range as at 2024 comprises:

– Changan BenBen E-Star: - A compact city car designed for urban commuting, featuring a modern design and efficient electric drivetrain.

- Changan Eado EV: - A sedan that combines style with electric efficiency, offering a comfortable and spacious interior along with a reasonable driving range.
- Changan CS15 EV: - An electric version of the popular CS15 compact SUV, providing a higher driving position and additional cargo space.
- Changan CS55 E-Rock: - An electric compact SUV known for its robust design, advanced features, and competitive range, claimed to be suitable for both city driving and longer trips.
- Changan UNI-K iDD: - A high-end electric SUV offering a blend of luxury, technology, and performance, targeting consumers looking for premium features in an electric vehicle.
- Changan Avatr 11: - Developed under the Avatr brand in collaboration with Huawei and CATL, this model represents Changan's push into the high-tech, smart electric vehicle segment with advanced connectivity and autonomous driving features.
- Changan Lumin: - A stylish and compact electric hatchback aimed at urban drivers, featuring a sleek design and efficient powertrain for daily commuting.

  In summary, Changan's current BEV lineup reflects its commitment to providing a diverse range of electric vehicles to meet the needs of different market segments, from compact city cars to premium SUVs. With continued advancements in technology and strategic collaborations, Changan is well-positioned to enhance its presence in the electric vehicle market. (source: Chat GPT).

BAIC

In 2009, the BAIC Group established a subsidiary, Beijing Electric Vehicle Co. Ltd. (BJEV) which became the first independently operated electric vehicle manufacture of China, and the first enterprise in China to be licenced for the production of 'new energy vehicles'. BAIC BluePark was subsequently restructured from its current subsidiary, Beijing Electric Vehicle Co. Ltd. (BAIC BJEV). In 2021, BAIC BluePark and Huawei made a joint statement that two companies would collaborate on building intelligent electric vehicle using its Arcfox brand. In 2022, BAIC BluePark launched the Arcfox α-S HI version under the 'Huawei Inside' mode. Huawei empowers the vehicle intelligence through the supply of both software and hardware. A crossover SUV, it is powered by a pair of electric motors that can produce 218 hp. (163 kW) and 265 lb.·ft. of torque by each motor, or combined 436 hp. (325 kW) and 530 lb.·ft. The motors are powered by a 93.6 kWh battery pack supplied by SK Innovation in South Korea. The crossover SUV claims a range 653 km (406 miles).

The ArcFox α-T comes standard with Level 2 autonomous driving and is equipped with 5G-compatible systems required to eventually meet Level 3 self-driving software updates. Some 6000 units were sold in 2021, but I have not been able to establish more recent sales information.

When the Beijing Electric Vehicle Co. Ltd. (BAIC BJEV) was reconsolidated into BAIC BluePark, the battery electric vehicle of Beijing brand, such as Beijing EU5, Beijing EU7, also came under the management of BAIC BluePark.

Current Beijing (EV models):

Beijing EU7 (2019–present), mid-size car, EV variant of Beijing U7
Beijing EU5 (2018–present), compact car, EV variant of Beijing U5
Beijing EX5 (2019–present), compact SUV, EV variant of Beijing X5

In March 2024, BAIC BluePark again restructured. The company is still indirectly controlled by BAIC Group, and the actual controller of the company is still the State-owned Assets Supervision and Administration Commission of Beijing Municipal Government.

Chery Auto

Chery Automobile Co. Ltd. was founded in 1997 in Wuhu, Anhui Province, to boost the local economy and produce affordable cars for the Chinese market. At first, it made vehicles under a production licence through a joint venture with SAIC Motor. It expanded quickly.

In 2012, Chery launched its first electric vehicle, the Chery QQ3 EV, marking its entry into the new energy vehicle (NEV) market selling around 1000 BEVs in 2012, 2000 in 2013, and about 3500 units in 2014.

In 2015, Chery sold approximately 5000 BEVs, expanding its BEV lineup with additional models catering to urban commuters.

In 2016, around 7000 units were sold, and in 2017, about 10,000 units, assisted by the launch of the Q1 name in line with the Chery eQ. The Chery eQ1 was powered by a permanent magnet synchronous electric motor of 30 kW (40 hp) with a top speed of 100 kmh (62 mph).

In 2017, Chery introduced the Arrizo 5e, an electric version of its popular Arrizo 5 sedan. This, together with the popularity of the EQ1 led to around 15,000 BEV sales in 2018.

The eQ was upgraded to the Eq5, with improved battery performance, and BEV sales increased to around 20,000 units in 2019, and around 25,000 units in 2020.

In 2021, this increased to about 30,000 units mainly with the Arrizo 5e contributing to higher sales; and in 2022, Chery sold some 40,000 BEVs.

2023 saw the introduction of a BEV version of Chery's Omodo 5 compact crossover SUV, with a power output of 150 kW (201 hp) and a 61 kWh battery which was introduced at the Auto Shanghai in April 2023. Named the Chery Omoda E5 or Omoda E5 depending on the market, the variant adopted a different front end styling along with other design changes. BEV sales increased to around 55,000 units in 2023.

Chery's commitment to developing and expanding its BEV lineup as adaptations of its mainstream successful models has resulted in a gradual but steady growth in

BEV sales over the years. The company's strong position in the Chinese automotive market, focus on technological advancements and market expansion, positions it well for future success in the electric vehicle market in the longer term.

Dongfeng

Dongfeng Motor Corporation, commonly known as Dongfeng (meaning 'East Wind' in Chinese), is one of the largest automobile manufacturers in China. It was established in 1969 under a directive from the Chinese government, and with state aid, to enhance the nation's automotive industry. Initially, it made trucks. Then with the commercial opening up of China, the rapid expansion of the Chinese car market in the 1990s and 2000s expanded into light vehicle production, primarily through joint ventures with western auto manufacturers who saw the massive market opportunity, such as the 1992 joint venture with Citroën (a subsidiary of PSA Peugeot Citroën) to form Dongfeng Citroën Automobile Company; the 2003 joint venture with Nissan to form Dongfeng Motor Co. Ltd.; the 2001 partnership with Honda to produce and sell Honda-branded vehicles in China; and the 2003 joint venture with Kia Motors to produce and sell Kia-branded vehicles. By the 2010s, Dongfeng had become one of China's largest automobile manufacturers, with a significant share of both the commercial and passenger vehicle markets.

Dongfeng Motor Corporation has been actively involved in the development and sales of BEVs as part of its broader strategy to embrace new energy solutions and contribute to environmental sustainability.

Dongfeng began exploring electric vehicle technology in the early 2000s, responding to the tenth revision of the 86/3 Program, but did not launch its first BEV until 2010 when Dongfeng Nissan launched the 'Venucia' brand, focusing on electric and hybrid vehicles. The Venucia e30, launched in 2014, was one of the first BEVs introduced under this brand and was designed to be compact and affordable for urban driving. It was based on, and shared most of the specifications of, the Nissan 'Leaf' (see above). It achieved somewhere around 500 units sold in 2014 and around 1000 units sold in 2015.

Then, under the Fengshen sub-brand, Dongfeng introduced the Fengshen A60 EV alongside the gasolene powered 2015 Fengshen A60 facelift in October 2015. The Fengshen A60 EV is based on the pre-facelift Fengshen A60 sedan and has a range of 200 km (120 miles) and a 130 kmh (81 mph) top speed with electric motor producing 95 hp. (71 kW). The Dongfeng Fengshen A60 EV was renamed to Dongfeng Fengshen E70 after the facelift placing the E70 inline with the 2015 facelift of the petrol-powered A60.

The group sold about 5000 BEVs in 2016.

In 2017, Dongfeng partnered with Renault and Nissan to form the 'eGT New Energy Automotive Co. Ltd.' to focus on the development and production of competitive BEVs for the Chinese market. In 2017, the group sold around 15,000 BEVs.

Several new BEV models were introduced between 2015 and 2020. The Dongfeng Fengshen E70 again received a facelift in 2019 and was revealed during the 2019 Shanghai Auto Show and now carried the nameplate of the revised English brand name, Aeolus, in the rear.

Other new models include the Dongfeng EX1, launched in 2019 as a compact electric SUV designed for urban commuting, based on the Renault City K-ZE blueprint, it was priced very competitively, its appeal being not in groundbreaking speed or dizzying technology, but in its humble commitment to making electric driving a norm for the many, not just the affluent few.

It also launched the Fengon E3, a crossover SUV BEV, the electric variant of the gasolene powered Fengon 500. The mechanics of the E3 are derived from the compact crossover Fengon ix5 but with a shorter wheelbase. Launched at the Shanghai 2019 motor show the E3 EV had a three-phase asynchronous electric motor delivering 120 kW combined with a 53.6 kWh lithium pherophosphate battery with a claimed range of over 400 km (250 miles).

In September 2022, Dongfeng officially unveiled the facelifted EV EX1 Pro in China with a rather low starting price of RMB 55,700 (just over US$ 7800). This is an updated version of the existing EV EX1, the major difference between the two models being a new ternary lithium battery pack. In reality, Dongfeng EV EX1 Pro is just a rebranded Renault City K-ZE, a Chinese version of the Dacia Spring, and they are all made in China or from Chinese parts (Dacia has no BEV assembly line). The car is manufactured by many other companies under different names and is actually based on another Renault—the Renault Kwid, a gasoline-powered small SUV.

Because of the complex nature of the company and its joint ventures gathering numbers on Dongfeng is difficult and must be considered no more accurate than 'in the region of', but BEV sales achievements for Dongfeng have been in the region of:

2018, around 30,000 units
2019, around 50,000 units
2020, around 75,000 units
2021, around 120,000 units
2022, around 180,000 units
2023, around 250,000 units (guestimate)

Dongfeng has also developed electric commercial vehicles, including buses and delivery trucks, catering to the growing demand for sustainable urban transportation solutions.

FAW

First Automobile Works (FAW) is one of China's oldest and largest automotive manufacturers. It was established in Changchun, Jilin province, in 1953. This was part of China's First Five-Year Plan to industrialise the country. FAW produced its first vehicle in 1956, the Jiefang CA10, a medium-duty truck modelled after the Soviet ZIS-150. It launched its first passenger car, the Hongqi (Red Flag) sedan, in 1973. In 1991, FAW entered into a joint venture with Volkswagen to form FAW-Volkswagen, which significantly boosted its passenger car production capabilities. This was followed in 1992 by another major joint venture with Toyota, named Tianjin FAW Toyota Motor Co. Ltd.; and in 2003, FAW and General Motors formed a joint venture to produce commercial vehicles. FAW produces a range of Volkswagen and Audi models for the Chinese market, one of the most successful joint ventures in the Chinese automotive industry.

In 2010, FAW established a new energy vehicle division, reflecting the company's commitment to sustainable development and green technology but did not produce BEVs in production quantities until 2017.

FAW launched the Besturn B30 EV, a BEV version of the B30 sedan in 2017, producing around 10,000 vehicles.

In 2018, FAW launched the Hongqi E-HS3, a BEV subcompact luxury SUV, under the prestigious Hongqi brand, the original Hongqi E-HS3 was launched in 2018 as a battery electric subcompact SUV. There are two versions of the powertrain, a front-wheel drive version with a single152 hp. electric motor and an all-wheel drive version with dual motors which achieve 304 hp. power output combined, FAW in total producing around 20,000 BEVs in 2018.

This was followed, in 2019, by the launch of the Hongqi E-HS9, a full-size luxury electric SUV. The E-HS9 featured advanced technologies and was designed to compete with high-end electric SUVs globally. The production Hongqi E-HS9 is a 5-door, 7-seat vehicle, and priced in the range of RMB 600,000 (US$ 80,000) to RMB 825,000 (US$ 110,000). The E-HS9 is available in two different performance variants. The lower-spec model features one electric motor for each axle rated at 215 hp. (160 kW) each, with 430 hp. (321 kW) combined. The top-trim model features a 329 hp. (245 kW) motor for the rear axle, with a combined power of 544 hp. (406 kW). The acceleration of the seven-passenger SUV from 0 to 60 mph (0–97 kmh) is within 5 s. Hongqi claim it has a range of approximately 300 miles (480 km). The E-HS9 has a 92.5-kilowatt-hour battery and the car supports wireless charging technology or non-contact charging, which can fully recharge the vehicle in 8.4 h and can adjust air suspension height for the best wireless charging alignment with charging efficiency up to 91%.

FAWs total BEV production in 2019 was around 35,000 vehicles.

The SiTech MEV is the second vehicle of SiTech, which came out in 2020. It is an electric vehicle based on the Bestune B30 sedan which has a range of 402 km and is powered by a 51 kWh battery. The MEV100 is a model jointly manufactured by SiTech and FAW. The SiTech MEV100 is essentially a rebadged Bestune B30 with the overall design based on the B30.

In 2020, the Bestune (formerly Besturn) brand introduced the E01, an electric SUV aimed at the mid-to-high-end market. This model featured modern design and competitive range, appealing to a broader audience, featuring a 190 hp. electric motor producing 236 lb-ft (320 Nm) of torque, a 61.34 kWh battery pack, and claimed a range of 450 km (279 miles).

FAWs BEV production in 2020 was in the range of 50,000 vehicles, and in 2021 increased to around 75,000 BEV units.

The introduction of the Hongqi E-QM5, at the latter stages of 2021, an electric sedan was aimed at providing a luxurious yet eco-friendly driving experience. The E-QM5 was designed to compete in both domestic and international markets and is equipped with a single electric motor that drives the front axle with 100 kW (134 hp), and a range of 431 km (268 miles) supported by a 55 kWh lithium iron phosphate battery. Reports suggest that later versions of the model could be equipped with 'Blade' batteries supplied by BYD. A battery-swap-enabled version called E-QH5 is also planned.

2022 also saw the introduction of the Bestune E05 (first shown at the 2021 Haikou International New Energy Vehicle and Connected Mobility Show). It has a 190 horsepower (142 kW) motor and a range of 450 km (280 miles) and features a sliding door on the right side to free passengers from opening the door and potentially hitting cyclists or pedestrians.

Four different versions are planned with two versions designated for taxis, a variant for the DiDi ride hailing service and a variant for the FAW-backed T3 ride hailing service.

In 2022,FAW turned out around 100,000 BEVs.

FAW continued to roll out upgrades in 2023 with improved battery life and range, raising its BEV production to around 150,000 units.

FAW's strategic focus on BEVs, supported by technological advancements and market expansion efforts, has positioned the company as a significant player in the electric vehicle market, both domestically and globally.

FAW plans to introduce more BEV models across its various brands, targeting different market segments from luxury to mass-market consumers.

GAC

Guangzhou Automobile Group Co. Ltd. (GAC Group) is a major Chinese automaker headquartered in Guangzhou, Guangdong province. The roots of GAC trace back to the establishment of Guangzhou Auto Repair Factory in 1954, which laid the groundwork for the future automotive manufacturing capabilities in the region. In 1985, Guangzhou Automobile Group was officially established as part of efforts to consolidate automotive manufacturing and enhance the industrial base in Guangzhou.

In 1997, GAC formed a joint venture with Honda Motor Co. Ltd., creating Guangzhou Honda Automobile Co. Ltd. This marked a significant step in GAC's growth, leveraging Honda's technology and expertise to produce passenger cars.

This was followed, in 2000, by another important joint venture with Toyota Motor Corporation, forming Guangzhou Toyota Motor Co. Ltd., further expanding its capabilities and market presence.

In 2014, GAC established a joint venture with Fiat Chrysler Automobiles (FCA) to produce Jeep-branded vehicles in China, expanding its portfolio with popular global models.

In 2017, GAC New Energy Automobile Co. Ltd. was established, reflecting the company's commitment to developing electric and hybrid vehicles. The Aion brand was launched to lead this new energy vehicle segment.

In 2018, GAC New Energy launched its first model, the Aion S, a BEV 4-seat compact sedan. The Aion S was launched at Auto Guangzhou in November 2018. It went on pre-sale in March 2019 and officially on sale in May 2019 powered by a 184 horsepower (137 kW) motor with a claimed range of 510 km (317 miles).

In 2019, GAC introduced the Aion LX, a high-end electric mid-sized SUV, with ADAS technologies and a number of long-range options: the LX 60 offered 503 km (313 miles); the LX 70, 520 km (323 miles); the LX 80, 650 km (404 miles); the LX 80D and LX 80D Max, 600 km (373 miles).

In 2019, GAC sold about 30,000 BEVs.

In 2020, GAC continued to expand its BEV sales launching the Aion V, a compact crossover SUV powered by a 135 kW (181 hp) permanent magnet synchronous motor such as the LX, the Aion V offered three trim levels, V 60, V 70, and V 80 Max, which offered ranges of 400 km (249 miles), 530 km (329 miles), and 600 km (373 miles), respectively. The BEV uses a lithium-ion battery and features Huawei 5G and V2X communication systems.

In 2020, GAC produced around 45,000 BEVs.

In 2021, GAC launched the Aion S Plus powered by a single electric motor with 204 hp (152 kW) with a 62 kWh battery, offering an NEDC range of 510 km (317 miles).

In 2021, GAC produced around 123,000 BEVs.

In 2022, GAC launched the Aion LX Plus with a 1008 km (626 miles) range.

In 2022, GAC produced around 271,000 BEVs.

In 2023, GAC launched a premium priced 2-seat electric sports car, the Aion Hyper SSR.

In 2023, GAC produced around 502,000 BEVs.

Successful collaborations with major global automakers such as Honda, Toyota, and FCA have been crucial in GAC's growth, providing access to advanced technologies and expanding its product lineup. As can be seen in the figures above, although starting BEV production later than many of its rivals, it is now one of the largest producers of BEVs.

Geely

Geely was founded in 1986 as a refrigerator parts manufacturer, but in 1994 diversified into the rapidly expanding motorcycle sector and quickly became one of China's top motorcycle manufacturers, transitioning to the automotive industry in 1997, with the first Geely car rolling off the production line in 1998. The company's initial foray into the automobile sector focused on low-cost vehicles.

In 2001, Geely was the first private Chinese automaker to receive government approval to produce cars and using private sector raised capital expanded its product line, introducing several new models and enhancing its production capabilities. The company also started exporting vehicles to international markets, including the Middle East, Africa, and South America.

Along with other Chinese auto manufacturers, Geely began exploring the electric market during the early 2000s, focusing on developing the necessary technology and infrastructure to support electric mobility. But it was not until 2008 that Geely launched its first BEV concept vehicle.

In 2010, Geely acquired Volvo Cars from Ford which provided access to advanced technology and expertise, significantly boosting its capabilities in electric vehicle development.

Geely's ownership of Volvo Cars has facilitated significant investments in a start-up called 'Polestar', providing the financial backing and technological resources necessary for developing cutting-edge electric vehicles. Polestar is a high-performance electric car brand that originated as a performance and tuning division of Volvo Cars before becoming a standalone brand focused on electric vehicles (EVs). Geely, a Chinese automotive giant, has played a significant role in Polestar's development through its ownership of Volvo Cars and strategic investments.

Geely and Volvo's collaboration has allowed Polestar to leverage Volvo's safety technology, engineering expertise, and global distribution network, enhancing its competitiveness in the EV market.

NOTE: Volvo and Polestar production figures are reported separately in the section covering Volvo below. The Geely production estimates provided in this section exclude Volvo and Polestar.

In 2011, the company showcased the Geely EC7-EV, a BEV version of the Emgrand EC7, a 5-seat luxury sedan offering a range of around 150 km per charge at the Shanghai Auto Show, commercially launching the product in 2014.

In 2018, Geely launched the Emgrand GSe, an SUV BEV, expanding its lineup of electric vehicles and catering to the growing demand for EVs in the SUV segment. A variant of the Geely Emgrand GS crossover, powered by a single electric motor producing 163 horsepower (122 kW) with a 52–61 kWh (190–220 MJ) battery giving a range of 353–400 km (219–249 miles). The 0–100 km per hour (0–62 mph) acceleration taking 9.9 s.

In 2015, Geely had launched a new brand, Geometry, dedicated to electric vehicles. The first model, Geometry A, was developed based on the Geely Emgrand GL sedan and was introduced in 2019. The Geometry A offered a choice of two battery capacities, a 51.9 kWh and a longer range 61.9 kWh version with a claimed range of 500 km (310 miles) on a single recharge driving a 120 kW (161 bhp) motor, with acceleration to 100 kmh (62 mph) in 8.8 s.

In 2019, Geely produced around 14,000 BEVs (excluding Volvo and Polestar).

In 2020, Geely continued to expand its electric vehicle lineup with the launch of the Geometry C, a compact SUV BEV. The Geometry C was based on the Geely

Emgrand GS and also offered a choice of two battery capacities, a 53 kWh battery and a longer range 70 kWh providing ranges of 400 and 500 km (249 and 311 miles) with a 200 hp. (150 kW) motor driving the front wheels and has a top speed of 150 kmh (93 mph).

In 2020, Geely produced around 75,000 BEVs.

In 2021, Geely announced a strategic partnership with Baidu to develop smart electric vehicles, leveraging Baidu's expertise in artificial intelligence and automated driving technology.

In 2021, Geely produced around 74,000 BEVs.

In 2022 Geely's Zeekr brand, focusing on premium electric vehicles, introduced the Zeekr 001, a high-performance BEV upmarket shooting brake with a range of up to 700 km. The top of the trim variant of the 001 has electric motors on both the front and rear axle, putting out a combined output of 536 hp. (400 kW) which can propel the 001 from 0 to100 kmh (0–62 mph) in 3.8 s, with a top speed of over 200 kmh (124 mph). Zeekr claims that the 001 can run on a full charge for up to 700 km (435 mi), and its charging setup will give it 120 km (75 miles) worth of power in just 5 min of charging time. A special model limited to 1000 units will be sold in 2023 containing 140 kWh to allow up to 1000 km (621 miles) of range.

In 2022, Geely produced around 149,000 BEVs.

In 2023, Geely's electric vehicle sales saw significant growth, driven by increasing demand and the successful launch of new models under the Geometry and Zeekr brands.

In 2023, Geely launched a new BEV model, and upgraded several existing models, with a particular highlight being the introduction of the Galaxy E8, a mid-to-large size BEV sedan, featuring, it claims, a sleek design and advanced technology. It offers multiple battery options (62 kWh, 75.6 Wh, and 76 kWh) with ranges between 550 and 665 km (341–413 m). The vehicle supports 800 V fast charging, allowing it to achieve a 180-km range with just a 5-min charge.

In 2023, Geely produced around 354,000 BEVs.

By leveraging strategic acquisitions, partnerships, and continuous technological advancements, Geely has positioned itself as a significant player in the global electric vehicle market.

Aiways

Aiways' first vehicle, the U5, was unveiled at the 2018 Beijing Auto Show. The U5 is a compact SUV with a range of up to 400 km (250 miles) on a single charge. It went on sale in China in 2019 and in Europe in 2020. In 2021, Aiways launched its second vehicle, the U6. The U6 is a coupe SUV with a range of up to 430 km (267 miles) on a single charge. Aiways planned to launch a third vehicle, the M7, in 2023. The M7 is a large SUV with a range of up to 500 km (310 miles) on a single recharge but in mid-2023, the company's financial situation collapsed, which led to the suspension of production. The company decided to intensify its activities on foreign markets, withdrawing Aiways from the difficult Chinese market and preparing to rebuild its plans for the European market. At the time of writing, its future is uncertain.

Additionally, there are several BEV specialising R&D institutes established, including Beijing Institute of Technology, Tongji University, and Tsinghua University. In summary, China is the largest market for BEVs in the world, accounting for around 58% of global production of BEVs.

In 2023, the China Association of Automobile Manufacturers reported that in China's 26 million new vehicle car market, 9.05 million passenger electric vehicles had been sold, consisting 6.26 million BEVs (battery-only EVs) and 2.79 million PHEV (plug-in hybrid electric vehicles). Plug-in electric vehicle (BEV and PHEV) sales were 37% of the overall automotive sales in China in 2023, with BEVs and PHEVs having 25% and 12% market share, respectively. This is a significant increase from 2020, when plug-in electric vehicles accounted for only 6.3% of total sales.

2010–2020 Enter Nissan, Tesla, Volvo, and the Big Established Auto Manufacturers

The chapter 'A New Millennium' above, in describing electric vehicle development in the first decade of the twenty-first century, described the entry and involvement of Nissan, Tesla, Volvo, Peugeot-Citroen into the world of the BEV. The chapters before summarised the efforts of GM, Ford, Chrysler, and other big automakers before them. And I rely on those summaries rather than repeat myself regarding their entry and involvement in BEV technology.

But it is true to say that, other than a few little and limited range Citroens/ Peugeots, and a few thousand Tesla Roadsters, in 2010 there was little presence of BEVs in the automotive market. A lot of talk, but little presence.

The previous chapter has summarised the emergence of the Chinese BEV since 2010. This chapter concentrates on the emergence of BEVs among their leading global auto manufacturing competitors.

Nissan

The reader will recall from above Nissan being the first of the major multinational auto manufacturers to launch a mass-produced BEV in 2010. The Nissan 'Leaf' featured a 24 kWh battery with a range of around 73 miles (117 km) per charge. In 2010, Nissan sold 9674 units.

In 2011, Nissan sold 22,689 BEV units.

In 2012 Nissan sold 26,973 BEV units.

In 2013, the Leaf received a larger 30 kWh battery option, increasing its range to approximately 107 miles (172 km).

In 2013, Nissan sold 47,716 BEV units.

In 2014, Nissan launched the e-NV200, a version of the NV200 van, aimed at businesses needing a zero-emission commercial vehicle. It shared many components with the Leaf.

In 2014, Nissan sold 61,507 BEV units (Nissan Leaf).

In 2015, Nissan sold 43,651 BEV units (Nissan Leaf, e-NV200).

In 2016, Nissan sold 49,430 BEV units (Nissan Leaf, e-NV200).

In 2017, Nissan launched the second-generation Nissan Leaf. The second-generation Leaf featured significant improvements in range, design, and technology. It initially launched with a 40 kWh battery, providing a range of around 151 miles (243 km).

In 2017, Nissan sold 54,451 BEV units (Nissan Leaf, e-NV200).

In 2018, the e-NV200 was upgraded with a 40 kWh battery, extending its range to around 124 miles (200 km).

In 2018, Nissan sold 87,149 BEV units (Nissan Leaf, e-NV200).

In 2019, the Leaf Plus was introduced, featuring a 62 kWh battery and a range of up to 226 miles (364 km).

In 2019, Nissan sold 69,670 BEV units (Nissan Leaf, e-NV200).

In 2020, Nissan launched the Nissan Ariya (market availability began in 2021). The Ariya is a crossover SUV BEV. It comes with multiple battery options: 63 kWh and 87 kWh, with ranges up to 300 miles (483 km) for the larger battery and featuring a dual-motor all-wheel drive option.

In 2020, Nissan sold 62,235 BEV units (Nissan Leaf, e-NV200).

In 2021, Nissan sold 50,500 BEV units (Nissan Leaf, e-NV200, Ariya).

In 2022, Nissan launched the Nissan Sakura. Aimed primarily at the Japanese market, the Sakura is a compact 'kei' car designed for urban driving. It features a 20 kWh battery, providing a range of around 112 miles (180 km).

Also in 2022, Nissan launched the Nissan Townstar EV. A successor to the e-NV200, the Townstar EV is a compact van designed for businesses.

In 2022, Nissan sold 65,700 BEV units (Nissan Leaf, Ariya, Townstar EV).

In 2023, Nissan sold around 100,000 BEV units (Nissan Leaf, Ariya, Townstar EV, Sakura).

Nissan has been a proponent of the CHAdeMO fast-charging standard, allowing for quicker recharging times for their vehicles, and Nissan has developed an ecosystem around their EVs, including home charging solutions, vehicle-to-grid technology, and battery recycling programmes.

The Leaf is to date one of the best-selling BEVs globally, with over 500,000 units sold by 2023.

Volvo

Volvo's early BEV developments have been covered in the chapters above.

In March 2010, Zhejiang Geely Holding Group Co. Ltd. (Geely) announced its intention to acquire Volvo Cars from Ford Motor Company. The acquisition was finalised in August 2010, with Geely purchasing Volvo Cars for $1.8 billion. This was significantly less than the $6.45 billion Ford had paid for Volvo in 1999. The synergies between Geely and Volvo have been summarised in the section on Geely above. Suffice to say that, according to their publicity, Geely have allowed Volvo to operate independently, providing substantial financial investment while respecting Volvo's management autonomy and brand identity.

Under Geely's ownership, Volvo has launched a series of successful models, including the XC90, XC60, S90, and the V90. Volvo have committed to electrification, aiming for all new models introduced after 2019 to be electric or hybrid. This culminated in the launch of several plug-in hybrid and battery electric vehicles, such as the XC40 Recharge.

Volvo has progressively expanded its lineup of BEVs since committing to electrification. During 2010–2018, Volvo focused primarily on hybrid models and concept electric vehicles but did not release any production BEVs.

In 2019, Volvo launched its 'XC40 Recharge' (with market availability beginning in 2020), its first production BEV. Based on the popular XC40 compact SUV, it featured a 78 kWh battery, offering a range of around 249 miles (400 km) on a single charge, driving dual electric motors providing all-wheel drive, with advanced safety features, and an integrated Google Android Automotive operating system.

Volvo sold just 6000 BEV units in 2020.

In 2021, Volvo launched its 'C40 Recharge'. Essentially the same car with a different body shape. This model is a coupe-SUV.

Volvo sold about 25,000 BEV units in 2021.

Volvo sold about 40,000 BEV units in 2022.

In 2023, Volvo launched its 'EX90'. The EX90 is a full-size, 7-seater SUV and represents Volvo's flagship BEV model, offering, Volvo claims, advanced driving features and a large battery for extended range, a luxurious interior, and high performance with an emphasis on sustainability.

Volvo sold about 60,000 BEV units in 2023.

Volvo announced its intention to become a fully electric car brand by 2030, underscoring its commitment to sustainability and reducing carbon emissions.

But Volvo also has a close relationship with Polestar. The relationship between Volvo and Polestar has evolved significantly over the years, transitioning from a motorsport collaboration to ownership and a full-fledged partnership in the development of high-performance and electric vehicles.

In 1996, Polestar was founded as Flash Engineering by Jan 'Flash' Nilsson in 1996. The team competed in the Swedish Touring Car Championship and was rebranded as Polestar Racing in 2005. The team became known for its success in motorsport, particularly with Volvo.

In 2009, Polestar began collaborating with Volvo to develop high-performance versions of Volvo's production cars. This partnership led to the creation of Polestar-branded performance models. In 2015, Volvo acquired Polestar, making it the official performance arm of Volvo cars. Polestar continued to produce high-performance versions of Volvo models under the Polestar brand.

In 2017, Volvo and its parent company Geely announced the transformation of Polestar into a standalone brand focused on electric performance vehicles. This marked a significant shift from Polestar's previous role as a performance tuner for Volvo.

In 2017, Polestar 1 was unveiled as a high-performance plug-in hybrid (PHEV) coupe with a limited electric-only range.

In 2020, Polestar 2, the brand's first BEV, was launched. It featured a fastback design and was aimed at competing with other premium electric vehicles such as the Tesla Model 3. The Polestar 2 is a BEV 5-door liftback and based on the CMA platform, but manufactured at the Geely Luqiao CMA Super Factory in Luqiao, Zhejiang, China. Production began in March 2020. With a combined twin motor of 300 kW (400 hp) powered by one of two battery pack options of 78 kWh (long range) offering 470 km (292 miles) or 72 kWh offering 450 km (280 miles) for the Chinese market.

In 2020, Polestar produced around 6000 BEV units.

In 2021, Polestar produced around 20,000 BEV units.

In 2022, Polestar announced the 'Polestar 3' SUV BEV. The Polestar 3 was officially revealed on 12 October 2022 is based on the same platform as the Volvo EX90 flagship SUV, both of which are based on the 2021 Volvo Recharge concept. With 2× Permanent Magnet Synchronous motors offering 489 hp. (365 kW) or 517 hp. (386 kW) (with Performance Pack) powered by a 111 kWh battery claiming a range of 610 km (380 miles).

In 2022, Polestar produced around 50,000 BEV units.

In 2023, Polestar produced around 100,000 BEV units.

The Polestar Precept, a concept car, first shown in 2020 but delayed by Covid, which highlighted the brand's future design direction and commitment to sustainability, will go into production in 2024 as the Polestar 5.

The Polestar 3 and upcoming Polestar 4 and 5 models aim to strengthen Polestar's presence in the electric vehicle market.

Polestar and Volvo share technologies and platforms, particularly in the areas of electric powertrains and vehicle architecture. The Compact Modular Architecture (CMA) and Scalable Product Architecture (SPA) are examples of shared platforms used by both brands.

If you are being charitable you might conclude that while Volvo focuses on a broader market with an emphasis on safety, luxury, and sustainability, Polestar targets the high-performance electric vehicle segment, appealing to customers seeking advanced technology and sporty driving experiences. If you are feeling more cynical, you might conclude that, under Chinese ownership, Volvo makes Volvos in Sweden, while the Nordic badged and priced Polestar makes Volvos in China.

The relationship between Volvo and Polestar has evolved from a motorsport collaboration to a strategic partnership focused on electrification and performance. This evolution reflects the changing dynamics of the automotive industry and the growing importance of electric mobility. Today, Volvo and Polestar work together to advance their shared goals of innovation, sustainability, and performance, while maintaining distinct brand identity.

Similar to TESLA, Polestar has adopted a direct-to-consumer sales model, bypassing traditional dealership networks to provide a seamless and transparent buying experience, although in many countries Polestar vehicles are also supported through the Volvo dealership/maintenance network.

Tesla

Before considering the other major auto manufacturers, we must turn our attention to a start-up company. But what a start-up. In the chapter, 'A New Millennium' above, we followed the creation of Tesla in 2003 through to the launch of its 'Roadster' model in 2008 and its subsequent market success. The Roadster continued to be sold until 2012.

In 2010, Tesla sold about 2000 BEV units (Roadster).

In 2011, Tesla sold about 2500 BEV units (Roadster).

Launched in June 2012, the Model S was Tesla's first mass-produced electric car. It was a luxury 4-door sedan with impressive range and performance, gaining widespread acclaim and solidifying Tesla's reputation. It is by now famous/infamous 'Autopilot' system being introduced in 2014. The rear axle has a traditional open differential. Models with dual drive dual motors also have an open differential on the front axles. The front and rear axles have no mechanical linkage—with dual motors, the power distribution among them is controlled electronically. The original Model S battery was made of several thousand cylindrical cells (18,650 of them).

These cells use a graphite/silicon anode and a nickel-cobalt-aluminium cathode with an aqueous electrolyte and lithium ions as charge carriers. The battery capacity within the Model S has changed numerous times since its debut, ranging from 60 to 100 kWh. The base Model S 60 was released with 60 kWh battery capacity and used a 270 kW (362 hp) motor. The model 85 was rear-wheel drive with an 85 kWh battery with 278 kW (373 hp) motor. The Performance variant (P85) offered a three-phase, four-pole AC induction 310 kW (416 hp) motor.

In 2012, Tesla sold about 2650 BEV units (Roadster, Model S).

In 2013, Tesla sold about 22,500 BEV units (Model S).

In 2014, Tesla sold about 31,600 BEV units (Model S).

Launched in September 2015, the Model X was Tesla's first SUV; a 7-seat 5-door SUV BEV. Known for its falcon-wing doors, it offered a combination of performance, utility, and advanced technology. The powertrain is taken from the S with similar options.

In 2015, Tesla sold about 50,580 BEV units (Model S, Model X).

In 2016, Tesla sold about 76,230 BEV units (Model S, Model X).

Unveiled in July 2017, the Model 3 was a more affordable sedan aimed at the mass-market. It played a crucial role in increasing Tesla's production volume and market penetration, a mid-size 4-door sedan. Rear-motor, rear-wheel drive and dual-motor, all-wheel drive models; 57.5, 79 or 82 kWh lithium-ion (earlier versions were: 54, 60, 62, 75, 78.1 kWh) battery giving a range of 272 m (438 km) (rear-wheel drive); 363 m (584 km) (long-range rear-wheel drive); 341 m (549 km) (long-range all-wheel drive).

In 2017, Tesla sold about 103,020 BEV units (Model S, Model X, Model 3).

In 2018, Tesla sold about 245,240 BEV units (Model S, Model X, Model 3).

Revealed in March 2019, the Model Y was a compact SUV built on the Model 3 platform. It offered similar performance and range but with additional space and versatility. Sales commenced in 2020.

In 2019, Tesla sold about 367,500 BEV units (Model S, Model X, Model 3).

In 2020, Tesla sold about 499,550 BEV units (Model S, Model X, Model 3, Model Y).

In 2021, Tesla introduced the high-performance Model S Plaid variant, boasting impressive performance.

In 2021, Tesla sold about 936,200 BEV units (Model S, Model X, Model 3, Model Y).

In 2022, there was continued focus on Cybertruck development with production delays pushing back its release further, but Tesla Semi production began, focusing on commercial heavy-duty transport with electric power.

In 2022, Tesla sold about 1,314,000 BEV units (Model S, Model X, Model 3, Model Y).

In 2023, production of the Cybertruck began.

In 2023, Tesla sold an estimated 1,800,000 BEV units (Model S, Model X, Model 3, Model Y, Cybertruck).

There are few words to describe Tesla's growth more appropriate than 'spectacular', especially when you consider that it has been built on selling premium product at premium prices well, well above their gasoline competition, and noticeably more expensive than their hybrid equivalents.

Tesla vehicles have received mixed reliability ratings over the years. For instance, Consumer Reports has sometimes ranked Tesla models below average in reliability due to issues such as fit and finish, electronics, and other mechanical problems. Tesla often scores lower than many traditional automakers in J.D. Power's Initial Quality Study, which measures the number of problems experienced per 100 vehicles during the first 90 days of ownership. In long-term reliability studies, Tesla's performance has also been varied, with some reports citing above-average issues compared to other brands. But despite mixed reliability scores, Tesla often scores very high in owner satisfaction surveys. Many Tesla owners are highly satisfied with their cars' performance, technology, and brand image, which sometimes outweighs the reported reliability issues.

Tesla's reliability has been improving over time, but it still lags behind many established automakers, particularly those known for their long-term dependability.

However, the high levels of owner satisfaction suggest that many Tesla drivers are willing to overlook these issues due to the brand's innovative technology, performance, and environmental benefits. (source: Chat GPT).

How Tesla will fare against an onrush of lower priced high-performance Chinese competition is yet to be seen. But Tesla results to date make an important statement—if you make your BEV an attractive enough proposition, the affluent will buy it despite its shortcomings.

Stellantis

Stellantis N.V. is a multinational automotive manufacturing corporation formed on January 16, 2021, through the merger of Fiat Chrysler Automobiles (FCA) and the PSA Group (Peugeot Sociét́é Anonyme). The merger created one of the world's largest automakers, combining a diverse array of brands from both parent companies. Stellantis operates in more than 130 countries and has manufacturing facilities in over 30 countries, making it a significant global player in the automotive industry.

Stellantis encompasses a wide range of well-known automotive brands, including: Abarth, Alfa Romeo, Chrysler, Citroën, Dodge, DS Automobiles, Fiat, Jeep, Lancia, Maserati, Opel, Peugeot, Ram Trucks, Vauxhall.

At the time of the creation of Stellantis in January 2021, several BEV models were already in production under the brands that became part of Stellantis. The primary BEV models that were in production at the time of the merger were:

- Peugeot e-208: A compact electric hatchback that offers a combination of style, performance, and electric efficiency.
- Peugeot e-2008: An electric version of the Peugeot 2008 SUV, providing an eco-friendly option in the popular compact SUV segment.
- Citroën ë-C4: An electric compact crossover that merges Citroën's distinctive design with zero-emission driving.
- Citroën Ami: A small, 2-seater urban electric vehicle designed for short trips and city driving.
- Opel/Vauxhall Corsa-e: An electric version of the popular Corsa, offering a practical and efficient urban mobility solution.
- Opel/Vauxhall Mokka-e: An electric variant of the Mokka SUV, available under both the Opel and Vauxhall brands.
- Fiat 500 Electric: An electric version of the iconic Fiat 500, providing a stylish and compact option for urban driving.
- DS 3 Crossback E-Tense: A luxury compact electric SUV that combines premium features with electric efficiency.

Commercial Vehicles: Peugeot e-Expert/Citroën ë-Jumpy/Opel/Vauxhall Vivaro-e: Electric versions of the mid-size vans, catering to commercial users looking for sustainable transport solutions.

These models reflect the initial efforts of the respective brands towards electrification and zero-emission driving, laying the foundation for Stellantis's broader strategy to expand its electric vehicle lineup post-merger. Apart from the BEV Ami, its BEV offerings to that date being electric variants of its combustion engine mainstream models.

Stellantis has been actively expanding its portfolio of BEVs since its formation in 2021. Here is a summary of the BEV models launched by Stellantis brands from 2021 to 2023:

2021

- Fiat New 500e: A modern electric version of the iconic Fiat 500, offering a compact and stylish urban mobility solution. The New 500 claims a 320 km (199 miles) range and is powered by an 87 kW (116 hp) electric motor, fed by a 42 kWh lithium-ion battery pack.
- Peugeot e-208: A compact electric hatchback that combines Peugeot's design with electric efficiency. The e-208 has a heat-pump controlled 50 kWh battery, a 100 kW (134 hp) motor, and a 6.6 kW charger. Claimed range is 340 km (211 miles).
- Opel/Vauxhall Mokka-e: An electric version of the Mokka SUV, available under both the Opel and Vauxhall brands, is powered by a 50 kWh lithium-ion battery with a range of 322 km (200 miles).
- Citroën ë-C4: An electric compact crossover combining Citroën's unique design with BEV driving. 156 hp. (115 kWh) motor with a claimed 415 km (260 m) range.

In 2021, Stellantis brands achieved about 120,000 BEV unit sales.

2022

- Peugeot e-Expert/Citroën ë-Jumpy/Opel/Vauxhall Vivaro-e: Electric versions of the mid-size vans, catering to commercial and business users looking for sustainable transport solutions.

In 2022, Stellantis brands achieved about 230,000 BEV unit sales.

2023

- Jeep Avenger EV: A fully electric SUV, the Avenger marks Jeep's entry into the BEV market with a focus on compact SUV efficiency and capability. A BEV version of a mainstream model it is equipped with a 400 V front-mounted elec-

tric motor manufactured by Emotors, producing 154 hp.; (115 kW) and a 54 kWh battery, the Avenger EV offers a range of up to 400 km (250 miles) (550 km (340 miles) in urban cycle).
- Peugeot e-Rifter/Citroën ë-Berlingo/Opel/Vauxhall Combo-e Life: Electric versions of the multi-purpose vans, designed for both passenger and light commercial use.
- DS 4 E-Tense: A luxury compact 2-door performance hatchback variant of the DS 4 from DS Automobiles, offering premium features and electric performance. Combustion engine, hybrid, and BEV versions available 402 hp. (300 kW) Premium version 815 hp. (608 kW).
- Maserati Grecale Folgore: Maserati's first fully electric compact SUV, a BEV version of its internal combustion model, aims to combine luxury and high performance with an electric powertrain. It is equipped with a 105 kWh battery and went on sale in 2023 as a 2024 model. The Folgore features dual motors that each produce 279 hp. (208 kW; 283), adding up to 558 hp. (416 kW) altogether. It claims a range of 500 km (310 miles).

In 2023, Stellantis brands achieved an estimated to be about 350,000 BEV unit sales.

These launches are part of Stellantis's broader strategy to offer more than 30 BEV models by 2030 and to make significant progress towards carbon neutrality.

Mercedes Group

Mercedes-Benz has been actively developing and launching BEVs under its various sub-brands since 2010.

In 2010, Mercedes-Benz launched the 'A-Class E-Cell' an early electric offering, based on the A-Class platform, with limited production.

In 2012, Mercedes owned 'Smart' launched the 'Smart Fortwo Electric Drive', an early BEV, focusing on urban mobility with a range of around 68 miles (109 km).

In 2013, estimated total BEV sales around 5000 units.

Mercedes-Benz subsequently consolidated its electric efforts under the EQ sub-brand, which stands for 'Electric Intelligence' but launched no new BEV models before 2014.

In 2014, Mercedes-Benz launched the 'B-Class Electric Drive', also known as the B250e; this compact BEV was based on the B-Class platform and featured electric drive technology developed in collaboration with Tesla.

In 2014, approximately 10,000 BEV units sold.

In 2015, around 12,000 BEV units sold.

In 2016, an estimated sale of about 15,000 BEV units.

In 2017, Smart EQ Fortwo and Forfour launched updated versions of the electric Fortwo and the introduction of the Forfour, with ranges of approximately 70–80 miles (113–129 km).

In 2017, approximately 20,000 BEV units sold, primarily from the Smart brand.

In 2018, Mercedes-Benz launched the 'eVito', an electric version of the Vito van, designed for urban delivery and transport, with a range of around 93 miles (150 km).

In 2018 around 25,000 BEV units sold, with the eVito and other models starting to contribute.

The next Mercedes BEV launch was in 2019 was the 'Mercedes-Benz EQC'. The EQC 400 4MATIC: The first SUV BEV under the EQ brand, featuring a range of approximately 200–220 miles (320–354 km) on a full charge, depending on the testing cycle.

In 2019, Mercedes-Benz also launched the 'eSprinter', an electric version of the Sprinter van, offering a range suitable for urban deliveries, around 93 miles (150 km) with the standard battery.

In 2019, Mercedes Group sold approximately 40,000 BEV units, driven by the launch of the EQC and increased sales of Smart BEVs.

In 2020, Mercedes launched the 'Mercedes-Benz EQV', an electric van based on the V-Class, designed for both passenger and cargo applications, with a range of around 250 miles (402 km).

In 2020, Mercedes Group achieved around BEV 50,000 unit sales, with the EQC and Smart BEVs continuing to be the main contributors.

In 2021, Mercedes-Benz launched its 'EQA', a compact SUV BEV positioned as the electric counterpart to the GLA, with a range of approximately 265 miles (426 km) in the EQA 250 variant.

Also in 2021 Mercedes-Benz launched its 'EQB', a larger compact SUV BEV, offering seating for up to seven passengers, based on the GLB platform, with a range of around 260 miles (418 km).

And also in 2021 Mercedes-Benz launched its 'EQS', a flagship sedan BEV, representing the electric equivalent of the S-Class, with a range of up to 478 miles (770 km) in the EQS. 450+ variant.

In 2021, it is estimated that Mercedes Group achieved around 100,000 BEV unit sales, with significant contributions from the new EQ models.

In 2022, Mercedes-Benz launched the 'EQE', a mid-sized sedan BEV. Positioned below the EQS, akin to the E-Class, with a range of up to 410 miles (660 km).

In 2022, Mercedes-Benz launched the 'EQS SUV', the SUV BEV counterpart to the EQS sedan, offering similar luxury features and technology, with a range of around 373 miles (600 km).

In 2022, Mercedes Group BEV unit sales expanded rapidly to approximately 150,000 units, driven by the popularity of the EQS and the introduction of the EQE and EQB.

In 2023, Mercedes-Benz launched the 'EQT', an electric small van, targeting both family and commercial use, derived from the T-Class.

In 2023, Mercedes Group BEV unit sales are estimated to exceed 200,000 units, with a full year of sales for the expanded EQ lineup including the EQS SUV and EQT.

From 2010 to 2020, Mercedes Group BEV expanded only steadily, but between 2020 and 2023, Mercedes-Benz has significantly expanded its portfolio of BEVs, covering various segments from compact cars and luxury sedans to SUVs, vans, and commercial vehicles. The EQ brand has been central to this strategy, showcasing Mercedes-Benz's commitment to an electric future. But, according to Auto Express (2024), Mercedes-Benz is to revert its electric cars back to traditional model naming structures in the coming years as the first generation of its all-electric models come to the end of their life cycles.

The new electric G-Class marks the change in approach. Branded 'G580 with EQ Technology', rather than the previously touted Mercedes EQG that would follow the naming strategy, the electric G-Class sits alongside petrol, diesel, and high-performance AMG models in one range of cars.

Volkswagen

Volkswagen Group (comprising Volkswagen, Audi, Porsche, Škoda, SEAT/Cupra, Bentley) has been actively developing and launching BEVs under various brands since 2011. The following list provides summary of the BEV models launched by Volkswagen Group from 2011 to 2023 and approximate VW Group sales for each year:

2013 - Volkswagen e-Up!: - A fully electric version of the Up! city car, with a range of around 99 miles (160 km).

In 2013, VW Group achieved BEV sales of less than 1000 units.

2014 - Volkswagen e-Golf: - An electric version of the popular Golf, with a range initially around 83 miles (134 km), later increased to approximately 125 miles (201 km) in updated versions.

In 2014, VW Group achieved BEV sales of around 1200 units.

In 2015, VW Group achieved BEV sales of around 2050 units.

In 2016, VW Group achieved BEV sales of around 3600 units.

In 2017, VW Group achieved BEV sales of around 5000 units.

2018 - Audi e-tron: - A luxury SUV BEV with a range of around 222 miles (357 km).

In 2018, VW Group achieved BEV sales of around 12,500 units.

2019 - Škoda Citigo-e iV: - A BEV version of the Citigo city car, with a range of around 170 miles (274 km).

2019 - Porsche Taycan: - A high-performance sedan BEV with various models (Taycan 4S, Turbo, Turbo S) and ranges between 200 and 300 miles (322–483 km).

In 2019, VW Group achieved BEV sales of around 29,000 units.

2020 - Volkswagen ID.3: - The first model based on the MEB platform, designed specifically for electric vehicles, with ranges varying from 205 to 342 miles (330–550 km), depending on the battery size.

2020 - Škoda Enyaq iV: - An SUV BEV based on the MEB platform, with ranges up to 316 miles (508 km).

2020 - Audi e-tron Sportback: - A sportier, coupe-style version of the e-tron SUV BEV, with a similar range of around 222 miles (357 km).

2020 - SEAT Mii Electric: - A BEV version of the Mii city car, with a range of around 160 miles (257 km).

In 2020, VW Group achieved BEV sales of around 57,000 units.

2021 - Cupra Born: - A sporty BEV hatchback based on the Volkswagen ID.3, with ranges up to 335 miles (539 km).

2021 - Volkswagen ID.4: - A compact SUV BEV with ranges between 250 and 323 miles (402–520 km), depending on the battery and drive configuration.

2021 - Audi Q4 e-tron: - A compact SUV BEV based on the MEB platform, with ranges up to 323 miles (520 km).

2021 - Audi Q4 Sportback e-tron: - A coupe-SUV BEV version of the Q4 e-tron, with similar specifications.

2021 - Audi e-tron GT: - A high-performance sedan BEV developed in collaboration with Porsche, with ranges up to 238 miles (383 km).

2021 - Porsche Taycan Cross Turismo: - A more rugged, wagon-like version of the Taycan, with similar performance and range.

In 2021, VW Group achieved BEV sales of around 263,000 units.

2022 - Volkswagen ID.5: - A coupe-SUV BEV variant of the ID.4, offering similar range and performance.

2022 - Volkswagen ID. Buzz: - An electric van inspired by the classic VW Microbus, with modern features and an expected range of around 250 miles (402 km).

2022 - Audi Q5 e-tron: - A BEV version of the popular Q5 SUV, tailored for the Chinese market.

In 2022, VW Group achieved BEV sales of around 572,000 units.

2023 - Volkswagen ID.7: - A large sedan BEV designed for long-range driving, offering advanced technology and expected ranges of up to 435 miles (700 km).

In 2023, VW Group achieved BEV sales of around 736,000 units.

From 2011 to 2023, Volkswagen Group has significantly expanded its BEV portfolio across its various brands, leveraging shared platforms such as the MEB to offer a diverse range of electric vehicles. These models cater to different market segments, from compact city cars and family SUVs to luxury sedans and high-performance sports cars.

General Motors

General Motors (GM) Group BEV launches and BEV units sales achieved from 2011 to 2023 excluding SAIC joint venture models in China (already covered in SAIC above) were as follows:

2011- Chevrolet Volt (extended-range electric vehicle (EREV), sometimes categorised as a BEV due to its electric driving capabilities, but in reality a PHEV because it has a range extending small petrol engine to put charge into the battery).

In 2011, GM Group (excluding China) sold approximately 7700 PHEV units (Chevrolet Volt).

In 2012, GM Group (excluding China) sold approximately 23,000 PHEV units (Chevrolet Volt).

2013 - Chevrolet Spark EV: This BEV had a 97 kW (130 hp) motor and was initially powered by a 21.3 kWh lithium iron phosphate (LiFePO) battery, but in the 2015 model the chemistry was changed to lithium-ion (Li-ion) which resulted in a smaller capacity pack (19 kWh) while maintaining the same EV range, which was estimated at 82 miles (132 km).

In 2013, GM Group (excluding China) sold approximately 23,000 PHEV/BEV units 23,094 (almost entirely Chevrolet Volt).

In 2014, GM Group (excluding China) sold approximately 20,000 PHEV/BEV units (18,805 Chevrolet Volt, 1145 Spark EV).

In 2015, GM Group (excluding China) sold approximately 18,000 PHEV/BEV units (15,393 Chevrolet Volt, 2629 Spark EV).

2016 - Chevrolet Bolt EV/ Opel Ampera-e: subcompact front-wheel drive 5-door hatchback. 200 hp. (149 kW) permanent magnet motor/generator. Battery: 2016–2019: 60.0 kWh lithium-ion; (2020–2023: 66.0 kWh lithium-ion). Range (2017–2019) 238 miles (383 km).

In 2016, GM Group (excluding China) sold approximately 27,000 PHEV/BEV units (24,739 Volt, 2122 Spark EV).

In 2017, GM Group (excluding China) sold approximately 44,000 BEV/PHEV units (23,297 Bolt EV, 20,349 Volt).

In 2018, GM Group (excluding China) sold approximately 36,000 BEV/PHEV units (18,019 Bolt EV, 18,306 Volt).

In 2019, GM Group (excluding China) sold approximately 16,500 BEV units (Bolt EV).

In 2020, GM Group (excluding China) sold approximately 21,000 BEV units (Bolt EV).

In 2021, GM Group (excluding China) sold approximately 25,000 BEV units (Bolt EV).

2022–Cadillac Lyriq: The Cadillac Lyriq is a mid-size 5-door rear-wheel drive luxury crossover SUV BEV with an all-wheel drive option, powered by a permanent magnet synchronous AC electric motor and a 100.4 kWh battery with a claimed range of 314 miles (505 km), rear-wheel drive, and 307 miles (494 km) all-wheel drive.

2022–Chevrolet Bolt EUV: The Chevrolet Bolt EUV (electric utility vehicle) is a subcompact crossover SUV BEV with a 200 hp. (149 kW) permanent magnet motor powered by a 65.0 kWh lithium-ion battery claiming a range of 247 miles (398 km).

In 2022, GM Group (excluding China) sold approximately 38,000 BEV units (Bolt EV and EUV).

2023–Chevrolet Equinox EV: launched in 2023 for sale in 2024, a compact crossover 5-door SUV BEV powered by a permanent magnet motor of 213–288 hp (159–215 kW) and a Ultium lithium-ion battery it claims a range of 319 m (513 km). Front-wheel drive and all-wheel drive versions available.

2023–Chevrolet Blazer EV is a battery electric mid-size 5-door crossover SUV BEV with a permanent magnet motor and 85 kWh or 102 kWh Ultium lithium-ion battery. Available in front-wheel drive, rear-wheel drive, and all-wheel drive option with ranges between 279 and 324 miles (449–521 km).

2023–GMC Hummer EV SUV: The GMC Hummer EV (badged as HEV) is a line of heavy-duty vehicles and is offered in two variants: a pickup truck and an SUV.

2023–Cadillac Celestiq (production commenced in 2024). A top of the range full-size luxury 5-door liftback sedan powered by dual-motor, four-wheel drive using two electric traction motors, one each for the front and rear axle, delivering an estimated combined output of 600 hp. (450 kW), with an estimated acceleration of 0–60 mph (97 kmh) in 3.8 s. The motors draw from a high-voltage Ultium traction battery with a total capacity of 111 kWh.

In 2023, GM Group (excluding China) sales are estimated to be approximately 70,000 BEV units.

This list highlights GM's gradual move into the BEV market and shows them ramping up their efforts in the 2020s.

Ford Motor

An overview of the Ford Motor Company Group's BEV launches and sales from 2011 to 2023 shows:

In 2011, the Ford Focus Electric was launched in late 2011. The Focus Electric uses the same body as the gasoline powered third-generation Ford Focus. On introduction in 2011, it had a 23 kWh liquid-cooled lithium-ion battery pack, which delivered a range of 76 miles (122 km).

In 2012, the Ford Focus Electric sold approximately 685 units.

In 2013, the Ford Focus Electric sold approximately 1738 units.

In 2014, the Ford Focus Electric sold approximately 1964 units.

In 2015, the Ford Focus Electric sold approximately 1582 units.

In 2016, the Ford Focus Electric sold approximately 1486 units.

In 2017, the range of the Ford Focus Electric was extended by upgrading to a 33.5 kWh liquid-cooled lithium-ion battery pack which delivered a range of 115 miles (185 km) and in 2017 Ford sold approximately 1817 units.

In 2018, the Ford Focus Electric sold approximately 1263 units. Production ended.

In 2020, Ford launched the Ford Mustang Mach-E. The model was launched with two battery pack sizes and three power outputs. The entry-level rear-wheel drive version was offered with either a 68 kWh battery pack driving a 266-horsepower (198 kW) motor or an 88 kWh extended battery pack, driving a 290-horsepower (216 kW) motor. The battery packs claiming ranges of around 230 and 300 miles (370 and 480 km), respectively.

A dual-motor all-wheel drive version was also offered with either the same 68 kWh battery pack driving a 266-horsepower (198 kW) motor, or the 88 kWh extended battery pack driving a 346-horsepower (258 kW) motor, offering ranges of 211 and 270 miles (340 and 435 km), respectively.

An all-wheel drive GT trim was offered with the 88 kWh pack, producing 480 horsepower (358 kW), a targeted 0–60 mph (0–97 kmh) time of 3.8 s, and a targeted driving range of 250 miles (400 km).

The Mach-E GT Performance Edition was revealed on December 2, 2020. It was equipped with a 480-horsepower (358 kW) motor and claimed a range of 235 miles (378 km).

In 2020, the Ford Mustang Mach-E sold approximately 3000 units (initial deliveries began late in the year).

In 2021, Ford launched the Ford E-Transit (Electric van for commercial use) 68 kWh (usable capacity) battery and up to 126 miles (203 km) range.

In 2021, the Ford Mustang Mach-E sold approximately 27,140 units (the Ford E-Transit was not available until late in the year).

In 2022, Ford launched the Ford F-150 Lightning, a full-size pickup truck, four-wheel drive, dual permanent magnet motors of 452 hp. (337 kW) or 580 hp. (433 kW) with battery options of 98.0 or 131.0 kWh lithium-ion offering ranges of 240 miles (390 km) (standard range); 320 miles (510 km) (extended range); or 300 miles (480 km) (Platinum trim).

In 2022, Ford sold approximately 59,000 BEV units (Ford Mustang Mach-E: approximately 39,458 units; Ford E-Transit: approximately 6500 units; Ford F-150 Lightning: approximately 13,258 units).

In 2023, Ford launched the Ford Explorer EV, a 5-door compact crossover SUV BEV offering both rear-wheel drive and dual-motor, all-wheel drive. Built on the Volkswagen Group MEB platform, it offers a powertrain of a permanent magnet synchronous motor powered by either a 55 kWh or 82 kWh lithium-ion battery with claimed ranges of 346–522 km (215–324 miles).

In 2023, Ford sold approximately 85,000 BEV units (Ford Mustang Mach-E estimated to be around 45,000 units; Ford E-Transit estimated to be around 10,000 units; Ford F-150 Lightning estimated to be around 25,000 units; Ford Explorer EV estimated to be around 5000 units).

Ford BEV sales build-up has been late in comparison to many of its competitors, but is accelerating in recent years, with significant launches such as the Mustang Mach-E and the F-150 Lightning, indicating a strategic shift towards electrification in both consumer and commercial vehicle segments.

BMW Group

BMW Group has progressively expanded its electric vehicle (BEV) lineup from 2011 to 2023.

Initially launched as a pilot project in 2008, the MINI E was one of BMW's first ventures into electric vehicles. It was produced in limited numbers for field trials. Based on the MINI Cooper, it had a 35 kWh battery, a range of about 100 miles, and served as a testbed for future electric vehicle technology. It was not sold commercially.

In November 2013, BMW launched the BMW i3, an urban high-roof hatchback with an electric powertrain. It used rear-wheel drive via a single-speed transmission and an underfloor lithium-ion battery pack with an optional range-extending petrol engine (647 cc 25 kW range extender); the electric powertrain being a BMW synchronous permanent magnet motor 125 kW or 137 kW.

The range extender in the BMW i3 functions as a backup power source, generating electricity to recharge the battery and maintain the electric motor's operation. It ensures that the vehicle can continue to be driven even when the battery's state of charge is low, but it does not directly power the wheels. This approach allows the BMW i3 to retain its identity as a primarily electric vehicle while offering the flexibility of extended range when needed Extended Range Electric Vehicle (EREV).

In 2014, the first full year of BMW i3 sales, BMW BEV sales were approximately 16,000 units.

In 2015, the second year of BMW i3 sales, BMW BEV sales were approximately 24,000 units.

In 2016, continued sales of BMW i3, BMW BEV sales were approximately 25,500 units.

In 2017, continued sales of BMW i3, BMW BEV sales were approximately 31,000 units.

In 2018, BMW launched the BMW i3s, a sportier version of the BMW i3 with increased power and improved handling.

In 2018, BMW i3 sales continue, BMW BEV sales were approximately 34,800 units.

In 2019, BMW i3 and the introduction of the MINI Electric pilot programmes saw BMW BEV sales rise to approximately 42,000 units.

In 2020, BMW launched the BMW iX3, a BEV version of the BMW X3, equipped with the fifth-generation BMW eDrive technology.

Also in 2020, the MINI Cooper SE BEV was formally launched. The first mass-production all-electric MINI, based on the MINI Cooper 3-door hatch. Powertrain of 135 kW (181 hp) synchronous electric motor with a 32.6 kWh battery with a claimed range of 235 km (144 miles).

In 2020, BMW BEV sales were approximately 44,530 units (including MINI Cooper SE).

In 2021, BMW launched the BMW i4, a BEV Gran Coupe, compact executive 5-door liftback/fastback sedan combining sporty design BEV drive. Both rear-motor, rear-wheel drive (eDrive) or dual-motor, all-wheel drive (xDrive) versions, 200–400 kW (268–536 hp) with lithium-ion 83.9 kWh battery offering a claimed range of 580 km (360 miles).

Also in 2021, BMW launched the BMW iX, a BEV SAV (Sports Activity Vehicle), flagship technology carrier with long-range capability and luxurious features.

In 2021, BMW BEV Sales were approximately 103,855 units (including MINI Electric).

In 2022, BMW launched the BMW iX1, a BEV version of the compact BMW X1.

Also in 2022 BMW launched the BMW i7, a BEV luxury sedan, part of the BMW 7 luxury sedan series. Battery capacities are rated at 105.7 kWh and 101.7 kWh, respectively, providing a range of up to 625 km (388 miles). The singular motor eDrive50, which was introduced in May 2023 as the entry-level i7, features BMW's eDrive which produces an output of 355 kW (476 hp) with an estimated range of 520 km (323 miles). The xDrive60 features dual eDrive motors which produce a combined output of 355 kW (476 hp) and a range from 591 to 625 km (367–388 miles). The range-topping M70 xDrive model features dual eDrive motors that produce a combined output of 485 kW (650 hp).

In 2022, BMW BEV sales were approximately 215,755 units (including MINI Electric).

In 2023, BMW launched the BMW i5, a BEV version of the BMW 5 sedan series, offered in rear-motor rear-wheel drive and dual-motor all-wheel drive (xDrive) versions 84.3 kWh (81.2 kWh) offering a range between 277 and 362 miles (445–582 km) depending on the model variant.

In 2023, BMW BEV sales are estimated to have exceeded 300,000 units.

BMW's strategy has been to integrate electric drivetrains across various segments, from compact urban vehicles to luxury sedans and SUVs, leveraging both all-electric and plug-in hybrid technologies. The focus has been on sustainability, performance, and incorporating the latest innovations in electric mobility.

BMW Group have announced plans to introduce additional BEV models under the MINI brand, including a new generation of the MINI Countryman in an electric variant.

Jaguar Land Rover

Jaguar Land Rover (JLR) has entered the BEV market with a focus on luxury and performance.

From 2011 to 2017, Jaguar Land Rover was primarily involved in developing and conceptualising its electric vehicle strategy. No BEV models were launched.

In 2018, Jaguar launched the Jaguar I-PACE. The Jaguar I-PACE was JLR's first SUV BEV, featuring a 90 kWh battery pack, a range of up to 234 miles, and dual electric motors providing all-wheel drive. It was designed to combine luxury, performance, and practicality, with a focus on dynamic handling and advanced technology.

In 2023, Jaguar introduced an updated version of the I-PACE with improved software, slight design tweaks, and enhanced charging capabilities to keep it competitive in the growing BEV market.

A year-by-year overview of Jaguar Land Rover's BEV sales from 2018 to 2023, focus primarily on the Jaguar I-PACE, JLRs only production BEV.

2018 I-Pace sales of approximately 6893 units
2019 I-Pace sales of approximately 17,355 units
2020 I-Pace sales of approximately 16,457 units
2021 I-Pace sales of approximately 12,006 units
2022 I-Pace sales of approximately 10,000 units
2023 I-Pace sales estimated to be around 9000 units

Despite this modest level of BEV success, Jaguar Land Rover has announced ambitious plans to become an all-electric luxury brand by 2025, with multiple new BEV models expected to be launched in the coming years.

Toyota Group

Toyota Group, known for its pioneering efforts in hybrid technology with the Prius, has been relatively cautious in its approach to BEVs. However, the company has made significant strides in the BEV market from 2011 to 2023. Here is an overview of Toyota Group's BEV launches and sales during this period:

In 2012, Toyota made a limited launch (California only) of the Toyota RAV4 EV (Second Generation): This was a BEV version of the RAV4 SUV, available in limited quantities in California, featuring a Tesla-supplied battery pack and powertrain. It was not repeated elsewhere.

Between 2013 and 2019, Toyota focused more on hybrid and hydrogen fuel cell technology, such as the Mirai, rather than pure BEVs.

But, under the pressure of governments pursuing all electric vehicle policies, relented and started to introduce BEV options.

In 2020, Toyota upmarket brand Lexus launched the Lexus UX 300e, a compact SUV BEV with a 54.3 kWh battery and a range of around 196 miles.

In 2020, Toyota Group sold around 1000 BEV units
In 2021, Toyota Group sold around 5800 BEV units

In 2022, Toyota launched the Toyota bZ4X: The first model in the bZ (beyond Zero) series, the bZ4X is an SUV BEV co-developed with Subaru (related to Subaru Solterra). It features a range of approximately 250 miles with a 71.4 kWh battery.

And in December 2022 Lexus launched the Lexus RZ 450e: a luxury electric SUV from Lexus, built on the same platform as the Toyota bZ4X, offering advanced features and a premium driving experience.

In 2022, Toyota Group sold around 15,000 BEV units.

In 2023, to the Chinese market, Toyota launched the Toyota bZ3: The bZ3 is a sedan BEV variant of the bZ BEV series, designed for the Chinese market, featuring advanced battery technology and a focus on interior space and comfort.

In 2023, Toyota Group sold around 72,000 BEV units.

Toyota Group's initial foray into BEVs has been cautious, with limited releases such as the second-generation RAV4 EV. The significant push into the BEV market began in earnest in 2020 with the Lexus UX 300e, followed by the introduction of the bZ4X and Lexus RZ 450e. Toyota's (reluctant) commitment to electric mobility is evident with the launch of the bZ series, aiming to expand its electric vehicle offerings globally.

Honda Group

The Honda Group has very cautiously entered the BEV market with limited launches from 2011 to 2023. An overview of their key BEV launches and sales during this period is as follows:

In 2012, Honda lunched the Honda Fit EV: A BEV version of the Honda Fit (known as the Jazz in some markets), the Fit EV had a 20 kWh battery and a range of around 82 miles. It was released in limited numbers, primarily for leasing in the USA and Japan.

In 2012, Honda Group achieved sales of about 200 BEV units.
In 2013, Honda Group achieved sales of about 500 BEV units.
In 2014, Honda Group achieved sales of about 400 BEV units.
In 2015, Honda Group achieved no sales of BEV units.
In 2016, Honda Group achieved no sales of BEV units.

Honda did not release another BEV until 2017, when it launched the Honda Clarity Electric: Part of the Clarity series (which also includes a fuel cell and plug-in hybrid variants), the Clarity Electric featured a 25.5 kWh battery with a range of around 89 miles. It was available only in select markets in the USA.

In 2017, Honda Group achieved sales of about 1000 BEV units.
In 2018, Honda Group achieved sales of about 2000 BEV units.
In 2019, Honda Group achieved sales of about 1900 BEV units.

In 2020, Honda launched the Honda e: A subcompact/supermini 5-door hatchback urban BEV, the Honda e features a retro design, rear-motor, rear-wheel drive with a permanent magnet brushless motor offering 100–113 kW (134–152 hp), powered by a 35.5 kWh lithium-ion battery and a claimed range of 220 km (137 miles). It was targeted primarily at European and Japanese markets, focusing on city driving and ADAS features.

In 2020, Honda Group achieved sales of about 5800 BEV units.
In 2021, Honda Group achieved sales of about 9500 BEV units.
In 2022, Honda Group achieved sales of about 11,000 BEV units.
In 2023, Honda Group achieved sales of about 12,000 BEV units.

Announced in 2021, but not in production until March 2024, the Honda Prologue is designed to be a mid-size crossover 5-door SUV BEV available in front-motor, front-wheel drive and dual-motor, all-wheel drive versions. With a powertrain of permanent magnet synchronous AC with a power output: Single motor: 218 hp.

(163 kW); Dual motor: 288 hp. (215 kW) powered by an 85.0 kWh Ultium lithium-ion battery claiming a range of 296 miles (476 km) front-wheel drive and 273–281 miles (439–452 km) all-wheel drive. It was developed in collaboration with General Motors, using GM's Ultium battery technology. The Honda Prologue is expected to offer a competitive range and advanced features suitable for the North American market.

Announced in 2022, but not in production before 2024, the Acura ZDX is designed to be an SUV BEV under Honda's luxury brand, Acura. The ZDX will also utilise GM's Ultium platform has similar powertrain to the Prologue and is part of Honda's strategy to electrify its luxury offerings.

In total in the period 2011–2023, Honda Group have sold less than 45,000 BEVs, but have models coming on stream in 2024.

Hyundai/Kia Group

The Hyundai/Kia Group has been progressively expanding its BEV lineup from 2010 to 2023 with the objective of becoming a prime world leader in the development and production of BEVs. An overview of their BEV launches and BEV sales during this period follows:

In 2010, Hyundai launched the Hyundai BlueOn: Based on the Hyundai i10, the BlueOn was Hyundai's first BEV with a 16.4 kWh battery and a range of about 87 miles. It was primarily used for pilot testing in South Korea.

In 2011, Hyundai's subsidiary sister company launched the Kia Ray EV: A small city-oriented EV based on the gasoline-powered Kia Ray, featuring a 16.4 kWh battery and a range of around 86 miles. It was mainly used for government fleets in South Korea.

In 2014, Kia launched the Kia Soul EV: This was the first globally available BEV from Kia, featuring a 27 kWh battery with a range of about 93 miles. It was praised at the time for its spacious interior and practicality.

In 2014, the Hyundai Group achieved sales of approximately 1000 BEV units.

In 2015, the Hyundai Group achieved sales of approximately 3000 BEV units.

In 2016, Hyundai launched the Hyundai Ioniq Electric: Part of the Ioniq lineup (which includes hybrid and plug-in hybrid variants), the Ioniq electric initially featured a 28-kWh battery with a range of 124 miles. A later version increased the battery size to 38.3 kWh, extending the range to 170 miles.

In 2016, the Hyundai Group achieved sales of approximately 6000 BEV units.

In 2017, the Hyundai Group achieved sales of approximately 16,000 BEV units.

In 2018, Hyundai launched the Hyundai Kona Electric: An electric version of the popular Kona SUV, available with a 39.2-kWh battery (189 miles range) and a 64-kWh battery (258 miles range). It gained popularity for its range, performance, and affordability.

Also launched in 2018 was the Kia Niro EV: A BEV version of the Niro crossover, sharing the same battery options as the Kona Electric (39.2 kWh and 64 kWh). It offered a similar range and was praised for its practicality and spacious interior.

In 2018, the Hyundai Group achieved sales of approximately 37,000 BEV units.

In 2019, the Hyundai Group achieved sales of approximately 87,000 BEV units.

In 2020, Kia launched the Kia Soul EV (Second Generation): The second generation featured a 64-kWh battery with a range of 243 miles, aligning it more closely with the Kona Electric and Niro EV.

In 2020, the Hyundai Group achieved sales of approximately 106,000 BEV units.

In 2021, Hyundai launched the Hyundai Ioniq 5: Built on Hyundai's new E-GMP platform, the Ioniq 5 featured retro-futuristic styling, 58 kWh and 77.4 kWh battery options, with ranges up to 300 miles. It included ADAS technology and fast-charging capabilities.

Also in 2021 Kia launched the Kia EV6: Also built on the E-GMP platform, the EV6 offered similar battery options to the Ioniq 5, with a range of up to 316 miles. It emphasised performance and design, positioning itself as a more sporty alternative.

In 2021, the Hyundai Group achieved sales of approximately 145,000 BEV units.

In 2022, Hyundai's luxury brand, Genesis, launched the Genesis GV60 a Compact Coupe 5-door luxury crossover SUV BEV: The first dedicated BEV from Genesis, the GV60 shared the E-GMP platform with the Ioniq 5 and 6, offering premium features and performance, offering rear-motor, rear-wheel drive and dual-motor, all-wheel drive using permanent magnet synchronous reluctance motors yielding 168–320 kW (225–429 hp) powered by a 77.4 kWh lithium-ion polymer battery with a claimed range of up to 516 km (321 miles).

Also in 2022 the Hyundai Ioniq 6 was launched: A mid-size 4-door fastback sedan based on the 'Prophecy' concept car, utilising the same E-GMP platform and featuring aerodynamic design and a focus on range and efficiency. Rear-motor, rear-wheel drive and dual-motor, all-wheel drive options using permanent magnet synchronous motors powered by a 53–77.4 kWh nickel-cobalt-manganese lithium battery giving a claimed range of up to 610 km (379 miles).

In 2022, the Hyundai Group achieved sales of approximately 234,000 BEV units.

In March 2023, Kia launched the Kia EV9: A large mid-sized crossover SUV BEV (concept revealed in 2021), becoming Kia's largest BEV. It also utilises the E-GMP platform, aiming for long range and large passenger capacity and has a three rows of seats option. 150 kW (201 hp) rear-motor, rear-wheel drive and dual-motor (additional 283 kW (380 hp)), all-wheel drive options; 76.1 kWh lithium-ion battery (Standard) and 99.8 kWh lithium-ion battery (Long Range) options offering a claimed range of up to 541 km (336 miles).

Also in 2023 the Genesis Electrified G80 (available in South Korea since the end of 2021) was launched globally: An electric version of the G80 luxury sedan, focusing on delivering a premium electric driving experience with advanced features and a substantial range. It is available in a dual-motor, all-wheel drive configuration. It uses two electric motors (272 kW combined), one at each axle, and an 87.2 kWh battery. The vehicle can recharge from 10% to 80% in 22 min, if a 350 kW charger is available. Just like the Ioniq 5, it has a vehicle-to-load (V2L) function, meaning that various appliances such as laptops or electric bikes can be plugged into the car (instead of a wall socket) and use the car as their power source. The Electrified G80 can be equipped with an optional solar roof.

In 2023, the Hyundai Group achieved estimated sales of approximately 332,000 BEV units.

The Hyundai/Kia Group has significantly expanded its BEV lineup from 2010 to 2023, transitioning from early pilot models such as the Hyundai BlueOn and Kia Ray EV to a diverse range of consumer-focused electric vehicles across their Hyundai, Kia, and Genesis brands. Their strategic use of the E-GMP platform has allowed them to offer competitive models with advanced technology, solid performance, and impressive range, reflecting their commitment to electric mobility. At the time of writing this work, outside of China, Hyundai Group, while still a long way behind Tesla and BYD, are at the forefront of vendors in respect of the volume of annual BEV units sold.

Summary: 2010–2025

Outside of China, in the period 2010–2019, several leading auto manufacturers continued to tinker with the development of BEVs, but mainly standing by and watching the development and growth of Tesla and developments in China. I think it is true to say that most of the leaders' attentions were focused on automated vehicles and 'monetising car data' promised by McKinsey as their prime way forward. It wasn't until the pressures exerted on them by governments reacting to their commitments made under the 2015 Paris Accord threatened their long-term business strategy, that they began to make serious investments in BEV technology. Some, eying Tesla and China, now seeing opportunity and grasping at it avidly or desperately, others, particularly the Japanese and US manufacturers, proceeding more cautiously.

Leading auto manufacturers had been heavily influenced by the promises being offered by developments in the automation of vehicle driving and the report, and long discussions leading up to its publication, 'Monetising Car Data- new service business opportunities to create new customer benefits' by McKinsey 2016 which advocated that 'As privately owned vehicles become increasingly connected to each other and to external infrastructures via a growing number of sensors, a massive amount of data is being generated. Gathering this data has become par for the course; leveraging insights from data in ways that can monetize it,….'

And.... 'How might industry players in the evolving automotive ecosystem turn car-generated data into valuable products and services?'

Suffice to say that it turns out that automated driving is far more complicated than just teaching the car when to start and stop, accelerate, brake, turn left and right. And monetising data more complicated than promised. But at the time, these issues were the focus of the minds of most auto manufacturers, rather than how to make the BEV a commercially attractive proposition……… that is until governments returned from making commitments under the Paris Accord and started to make demands on auto manufacturers that disrupted their existing long-term plans.

As a result, the period 2019–2023 has seen a ramp-up of investment in BEV technology by most, fervid development of BEVs by some, and a rapid but patchy take up of BEV sales in the marketplace.

So how did the 2015 Paris Accord change the game?

The Paris Agreement of 2015

The 2015 Paris Agreement is a landmark international treaty adopted on December 12, 2015, at the 21st Conference of the Parties (COP21) under the United Nations Framework Convention on Climate Change (UNFCCC) in Paris, France. It represents a global effort to combat climate change and limit global warming. The central aim is to keep the global average temperature increase well below 2 °C (3.6 °F) above pre-industrial levels, while pursuing efforts to limit the increase to 1.5 °C (2.7 °F). This is to significantly reduce the risks and impacts of climate change.

Each participating country must submit a plan, known as a Nationally Determined Contribution (NDC), outlining its climate actions, including emission reduction targets. Countries are expected to update and enhance these plans every 5 years to reflect increased ambition.

The Paris Agreement establishes a robust transparency framework, where countries have to regularly report their emissions and progress in implementing their NDCs. The reports are subject to international review. The Paris Agreement is legally binding, but the specific emission reduction commitments (NDCs) are not legally enforceable. Instead, the agreement relies on transparency, accountability, and peer pressure to encourage compliance.

The agreement aims to reach global peaking of greenhouse gas emissions as soon as possible and achieve a balance between anthropogenic emissions by sources and removals by sinks of greenhouse gases in the second half of the century, effectively moving towards net-zero.

Both the Kyoto Protocol and the Paris Agreement are about reducing the emissions that are causing global warming, but do not directly specify what those causes are. But as the Paris Agreement requires governments to specify the specific measures and actions they are taking (and be measured on the effectiveness of their actions), it is necessary to delve a little deeper and to identify the causes and scope for reduction, the Paris Agreement/UN rely on the IPCC.

IPCC Reports and Climate Action

The IPCC's Sixth Assessment Report (2021–2022) underscores the dire need for immediate and comprehensive action to mitigate climate impacts. In summary, it asserts, based on its 30 years of funded research that the link between CO_2 emissions and global warming is based on the scientific understanding of the greenhouse effect:

The Earth's atmosphere contains gases known as greenhouse gases (GHGs), which include carbon dioxide (CO_2), methane (CH_4), water vapour, and others. These gases have the ability to trap heat from the Sun. When sunlight reaches Earth, it is absorbed by the surface and then radiated back as infrared (heat) energy. Greenhouse gases absorb and re-radiate this heat, preventing it from escaping into space. This process keeps the Earth's surface warmer than it would be without these gases, making life possible.

Role of CO_2

CO_2 is one of the most significant greenhouse gases although it is not the most potent on a per-molecule basis (methane, for example, is more potent). However, CO_2 is more abundant and has a long atmospheric lifetime, making it a key driver of the greenhouse effect.

Human activities, particularly the burning of fossil fuels (coal, oil, and natural gas) for energy, deforestation, and industrial processes, have significantly increased the concentration of CO_2 in the atmosphere since the Industrial Revolution. The more CO_2 and other greenhouse gases that accumulate in the atmosphere, the more heat is trapped, leading to an overall warming of the planet's surface. Scientific measurements show a clear increase in global temperatures correlating closely with rising CO_2 levels. Over the past century, global average temperatures have risen by about 1.2 °C (2.2 °F), with most of this warming occurring in recent decades.

NOTE: There is also a strong correlation between the growth of the human population and the rate of global warming. More on this below.

Global warming is primarily driven by the increase in greenhouse gases such as CO_2, but several other factors and emissions contribute to the phenomenon. Here are some of the key causes of global warming beyond CO_2 emissions:

Methane (CH_4) Emissions

From several sources.

- Agriculture: Methane is released during the digestion of food by livestock, particularly cattle (enteric fermentation).
- Landfills: Decomposition of organic waste in landfills generates methane.
- Natural Gas and Oil Production: Methane leaks during the extraction, processing, and transportation of natural gas and oil.
- Wetlands: Natural wetlands are significant sources of methane due to anaerobic decomposition.

Methane is a potent greenhouse gas, with a global warming potential approximately 25 times greater than CO_2 over a 100-year period, though it remains in the atmosphere for a shorter time.

When methane (CH_4) is burned, it reacts with oxygen (O_2) in the air to produce carbon dioxide (CO_2), water (H_2O), and heat, thus adding to CO_2 emissions.

Nitrous Oxide (N_2O) Emissions

From several sources.

- Agriculture: The use of synthetic fertilisers, manure, and certain agricultural practices increase nitrous oxide emissions.
- Industrial Processes: Some chemical manufacturing processes produce nitrous oxide as a byproduct.
- Burning of Organic Matter: Biomass burning, including forest fires and agricultural burning, releases N_2O.

Nitrous oxide has a global warming potential about 298 times that of CO_2 over 100 years, making it a significant contributor despite its lower concentrations.

Fluorinated Gases

Of several types.

- Hydrofluorocarbons (HFCs): Used as refrigerants, in air conditioning, and in other industrial applications.
- Perfluorocarbons (PFCs): Used in electronics manufacturing and as solvents.
- Sulphur Hexafluoride (SF_6): Used as an insulator in the electrical industry.

These synthetic gases are extremely potent greenhouse gases, with global warming potentials thousands of times greater than CO_2. Although their concentrations are low, they are significant contributors due to their high potency and long atmospheric lifetimes.

Deforestation and Land Use Changes

When forests are cleared or burned for agriculture, urban development, or logging, the carbon stored in trees and soil is released as CO_2. Deforestation also reduces the number of trees available to absorb CO_2 from the atmosphere, exacerbating global warming.

Deforestation contributes significantly to global CO_2 emissions and disrupts local and global carbon cycles, increasing atmospheric CO_2 levels.

Black Carbon (Soot)

The burning of fossil fuels: Incomplete combustion of fossil fuels, especially coal and diesel, produces black carbon. Biomass burning: Forest fires, agricultural burning, and the burning of wood for cooking and heating.

Black carbon absorbs sunlight and heats the atmosphere. When it settles on ice and snow, it reduces their albedo (reflectivity), causing them to absorb more heat and melt faster, contributing to warming.

Ozone (O_3) as a Greenhouse Gas

Formed by reactions between pollutants such as nitrogen oxides (NOx) and volatile organic compounds (VOCs) in the presence of sunlight. Major sources include vehicle emissions, industrial processes, and biomass burning.

While stratospheric ozone protects life on Earth by absorbing harmful ultraviolet radiation, tropospheric (ground-level) ozone acts as a potent greenhouse gas, trapping heat in the atmosphere and contributing to global warming.

Water Vapour

Water vapour is the most abundant greenhouse gas, and its concentration is primarily controlled by natural processes. However, as the Earth warms, the atmosphere can hold more water vapour, which in turn amplifies warming in a positive feedback loop.

While not directly a cause of global warming, increased water vapour due to higher temperatures enhances the greenhouse effect, leading to further warming.

Changes in Land Use and Urbanisation

Urbanisation often involves replacing natural landscapes with concrete and asphalt, which absorb more heat than natural vegetation. This contributes to the urban heat island effect, where cities are warmer than surrounding rural areas, affecting local climates and contributing to global warming.

Land use changes, especially dense urban development, can alter local climates and contribute to the global increase in temperatures.

Natural Factors (with Limited Current Impact)

- Volcanic Activity: While volcanic eruptions can release large amounts of CO_2, they also emit aerosols that temporarily cool the Earth by reflecting sunlight. Over the long term, volcanoes have a minor impact compared to human activities. In the short term, significant eruptions can increase cloud cover, temporarily reducing access to sunlight, temporarily reducing global temperatures.
- Solar Variability: Changes in the Sun's energy output can influence the Earth's climate, but recent observations show that current global warming cannot be explained by changes in solar radiation alone.

Oceanic Emissions and Changes

Oceans release CO_2 and methane through natural processes, and changes in ocean temperatures can alter the amount of these gases released. Warming oceans also lead to decreased CO_2 absorption, reducing their role as a carbon sink.

These processes can contribute to the global greenhouse gas burden, through other natural oceanic processes.

Cyanobacteria

Cyanobacteria, also known as blue-green algae, have played a crucial role in Earth's history, particularly through their ability to photosynthesise and release oxygen. The following paragraphs show how their oxygen production can relate to global warming:

- **Historical impact of cyanobacteria:** Cyanobacteria were among the first organisms to perform oxygenic photosynthesis, releasing oxygen as a byproduct. Around 2.4 billion years ago, they contributed to the Great Oxygenation Event (GOE), which dramatically increased the amount of oxygen in Earth's atmosphere. It has only been recognised in recent years that cyanobacteria played an even greater role in this process than trees.
- **Water as an electron donor:** In the process of photosynthesis, cyanobacteria use sunlight to convert carbon dioxide (CO_2) and water (H_2O) into glucose (a type of sugar) and oxygen (O_2). The water provides electrons (through a process called photolysis), and oxygen is released as a byproduct.

- The simplified chemical equation for photosynthesis in cyanobacteria (and plants) is:

$$6CO_2 + 12H_2O + \text{Light Energy} \rightarrow C_6H_12O_6 + 6O_2 + 6H_2O$$

- **Oxygen release:** The oxygen that cyanobacteria release during photosynthesis comes from the splitting of water molecules. When cyanobacteria absorb sunlight, they use the energy to split water (H_2O) into oxygen (O_2), protons (H^+), and electrons (e^-). The oxygen is then released into the surrounding environment, which could be seawater or the atmosphere.
  The oxygen released by cyanobacteria (from both CO_2 but particularly the seawater it is living in) eventually led to the formation of the ozone layer, which protects the Earth from harmful ultraviolet (UV) radiation. This allowed more complex life forms to evolve.
- **Photosynthesis and carbon sequestration:** Through photosynthesis, cyanobacteria absorb carbon dioxide (CO_2) from the atmosphere and convert it into organic matter. This process reduces atmospheric CO_2 levels, which can mitigate global warming. Further, some cyanobacteria contribute to carbon sequestration by forming sediments that trap carbon over geological timescales. For instance, cyanobacteria in oceans can die and sink to the ocean floor, where their carbon content is stored away from the atmosphere.
  Oxygen itself does not have a direct warming or cooling effect. However, the oxygen-producing activities of cyanobacteria indirectly influence climate by removing CO_2 from the atmosphere, which reduces the greenhouse effect. Cyanobacteria help maintain a balance in the global carbon cycle. When they proliferate, they can lead to significant reductions in CO_2, which could theoretically counteract some aspects of global warming. There is interest in leveraging cyanobacteria or similar organisms for carbon sequestration as a form of geoengineering to combat climate change. However, such interventions would need careful management to avoid unintended ecological consequences.
  Most cyanobacteria thrive in warm water, and their 'blooming' proliferation noticeable from sea warming phenomena such as 'El Nino', increasing their carbon capture and oxygen liberation activity.
- **In summary**, while CO_2 emissions are seen as the primary driver of global warming, other gases such as methane, nitrous oxide, and fluorinated gases, as well as activities such as deforestation and black carbon emissions, also play significant roles in enhancing the greenhouse effect and contributing to the planet's warming. Cyanobacteria have an ameliorating effect.
  Cyanobacteria may have a role to play in the reduction of harmful emissions from vehicles, but more about that later in this work.

Source of Human-Generated CO_2 Emissions

Human-generated CO_2 emissions are generated from various sectors and activities. In order to identify their relative importance and significance, the following summaries (drawn from ChatGPT and Copilot) provide some very approximate proportions of the different human-generated sources of CO_2 emissions, along with approximate percentage points based on global data:

NOTE: These figures are very approximate and vary on a seasonal basis and from country to country and should be taken only as approximate indicators of the relative global proportionality between the different sources.

Energy Sector (Electricity and Heat Production): Percentage points: ~**42**%.

- Coal: The largest single source of CO_2 emissions within the energy sector, due to its high carbon content.
- Natural Gas: Emits less CO_2 than coal per unit of energy but is still a significant source.
- Oil: Used in electricity generation in some regions, contributing to CO_2 emissions.

Transportation: - Percentage points: ~**24**%.

- Road Transport: Cars, trucks, and buses running on gasoline and diesel are the main contributors.
- Aviation: Aircraft emissions are a smaller but growing part of transportation-related CO_2 emissions.
- Shipping: Maritime transport also contributes, primarily through the use of heavy fuel oil.

Industry: - Percentage points: ~**21**%.

- Cement Production: A major industrial source of CO_2, accounting for about 8% of global emissions on its own, due to the chemical process of calcination.
- Steel and Iron Production: Significant sources of CO_2 due to the use of coal in blast furnaces.
- Chemical Production: Includes emissions from the production of chemicals such as ammonia and fertilisers.

Buildings (Residential and Commercial): - Percentage points: ~**6**%.

- Heating and Cooling: CO_2 emissions from the combustion of fossil fuels for heating buildings, as well as electricity use for cooling.
- Cooking and Lighting: Emissions from the direct use of fossil fuels for cooking and from electricity consumption.

Agriculture, Forestry, and Other Land Use (AFOLU) - Percentage points: ~**11**%.

- Deforestation: CO_2 emissions from the burning or decomposition of trees and other vegetation during land clearance.
- Agricultural Practices: Emissions from the use of fossil fuels in machinery and from the production and use of fertilisers.
- Peatland Degradation: CO_2 released from drained peatlands, which are rich in carbon.

Waste Management: - Percentage points: ~**3**%.

- Landfills: Decomposition of organic waste in landfills produces CO_2 (in addition to methane).
- Incineration: Burning of waste materials, especially plastics, contributes to CO_2 emissions.

Transportation Components:

Transportation, at about 24%, is a significant source of global CO_2 emissions, and a legitimate target for significant CO_2 reduction, but various modes of transport contribute differently, therefore further analysis is needed. A comparative analysis of the main sources of transportation CO_2 emissions within the transportation sector, along with approximate percentage points yields the following:

Road Transport: - Percentage points: ~74%.

- **Passenger Cars:** Responsible for approximately 45% of transportation-related CO_2 emissions.
- **Light-Duty Trucks (including SUVs and vans):** Account for about 15%.
- **Heavy-Duty Trucks and Buses:** These vehicles contribute roughly 14%, due to their larger size, higher fuel consumption, and longer distances travelled.

Aviation: - Percentage points: ~12%.

- **Commercial Aviation:** The majority of aviation-related emissions come from commercial flights, both domestic and international.
- **General Aviation:** Smaller private planes and non-commercial flights contribute a minor share.

Shipping (Maritime Transport): - Percentage points: ~11%.

- **International Shipping:** The largest share of maritime emissions comes from large cargo ships that transport goods across the globe.
- **Domestic Shipping:** Coastal and inland waterway transport also contributes but to a lesser extent.

Rail Transport: - Percentage points: ~2%.

- **Freight Trains:** In some regions, freight trains are a significant source of rail-related CO_2 emissions, especially when diesel-powered or steam-powered.
- **Passenger Trains:** Emissions from passenger rail are typically lower, particularly in countries where electric trains predominate.

Rail transport is relatively more efficient, especially where electrification is prevalent, leading to its smaller share of emissions.

Other Transportation (Pipeline Transport, etc.): - Percentage points: ~1%.

- Pipelines: CO_2 emissions come from the energy used to pump oil, gas, and other materials through pipelines, usually driven by compressors powered by natural gas or electricity.

Key Takeaways.

- Power generation is by far the largest source of human-generated CO_2 emissions (at over 40%).
- Transportation is the second largest generator, of which Road Transport is the largest source of transportation-related CO_2 emission, of which some 15% is currently generated by cars and light commercial vehicles (the focus of this work). And so is a target for significant CO_2 emissions reductions.

Electricity and heats	40
Industry	20
Buildings	5
Waste management	2
Agriculture	10
Pipeline	1
Transport	24
Road	
Car	11
Light commercial	3
Heavy commercial/bus	3
Aviation	3
Shipping	2
Rail	1

Number of Cars or Number of Humans

There has been recent debate about whether global warming is affected more by the number of humans on the planet than a gas, CO_2, that constitutes only about 0.04% of the Earth's atmosphere by volume, and contention that the correlation between population growth and global warming is closer than that to CO_2 emissions. They are clearly all interrelated matters. This work is about BEVs, so the wrong place for such debate, but as it is closely related to the suitability of current political actions affecting BEV replacement of gasolene powered vehicles it would be remiss not to point out that despite the weight of arguments supporting the emissions theory, while the presence of global warming is now more or less uncontested, there are still those who offer evidence of other potential causes.

Here is what the IPCC have to say on the matter:

The Intergovernmental Panel on Climate Change (IPCC) addresses the relationship between population growth and global warming in various reports (summarised by ChatGPT).

IPCC: Population Growth as a Driver of Emissions

Increased Demand for Resources: Population growth leads to increased demand for energy, food, water, and other resources, which can result in higher greenhouse gas (GHG) emissions, especially when these needs are met through carbon-intensive means.

Urbanisation and Infrastructure Development: Growing populations often lead to expanded urban areas and infrastructure, increasing energy use and emissions from construction, transportation, and housing.

IPCC: Per Capita Emissions

Differences by Region: The IPCC highlights that the impact of population growth on global warming is not uniform across the world. In many high-income countries, per capita emissions are much higher than in low-income countries, even if population growth rates are lower.

Economic Growth and Consumption Patterns: Population growth interacts with economic growth and consumption patterns. Even with stable or declining populations, if per capita consumption and energy use increase, emissions can continue to rise.

IPCC: Population and Sustainability

Sustainable Development Goals (SDGs): The IPCC links population growth to broader sustainability issues. Managing population growth in a sustainable manner involves improving education, healthcare, and economic opportunities, particularly for women, which can lead to lower fertility rates and more sustainable consumption patterns.

Equity and Development: The IPCC emphasises the importance of addressing inequities in emissions and development. Reducing poverty and enhancing access to clean energy can help manage the environmental impacts of population growth.

IPCC: Mitigation Strategies

Technological and Behavioural Change: The IPCC suggests that technological innovations, such as clean energy technologies, and changes in behaviour, such as reducing meat consumption, can offset some of the impacts of population growth on emissions.

Policy Measures: Effective policies, including those that promote sustainable urbanisation, energy efficiency, and renewable energy, can help mitigate the effects of population growth on global warming.

IPCC: Long-Term Projections

Scenarios and Pathways: The IPCC uses scenarios to project future emissions and climate impacts, taking into account different assumptions about population growth, economic development, and technological change. In scenarios where population growth is combined with high levels of consumption and fossil fuel use, emissions are much higher than in scenarios with lower population growth and sustainable development practices.

IPCC: Summary

The IPCC acknowledges that population growth is a significant factor in global warming, but it is one part of a complex web of factors including economic development, consumption patterns, technology, and policy choices. Addressing global warming requires a holistic approach that considers population dynamics alongside efforts to reduce emissions through technological, behavioural, and policy interventions.

The Darwinian Outcomes

A vehicle owner/user selects the solution that 'best' meets their needs within the constraints pertaining at the time. And the best solution is not a common ideal, but varies from driver to driver, and changes over time.

The historical summaries given in the early chapters above are not simply there as a record of events, but as a demonstration of these choices and their consequences.

Steam, rails, and canals presented the opportunity to haul larger loads. Electricity didn't need time to get up a head of steam, nor a crankshaft to start, was quieter and less smelly than gasoline engines, but once those issues were ameliorated, faced bigger refuelling issues (such as availability, safe, and efficient storage and transfer). Rails, overhead cables, and canals required more upfront infrastructure, forward investment, and maintenance than the diesel bus. But in a different set of user circumstances there is a viable use case for mains power or rechargeable batteries.

So as long there is a viable use case, the technology will survive and most likely evolve. But what is a 'viable' use case?—it has to realistically meet the requirements of a class of user (including making economic sense). Thus, in the first third of the twentieth century, electricity offered a viable solution for public transport in cities, but failed for personal transport.

And when the use case ceases to be viable, the technical solution, in Darwinesque fashion, is doomed, like the 'fardier à vapeur', to take its place in a museum.

Chapter 4
Consequential Impacts of the Current BEV

Abstract Chapter 4 considers the consequential impacts of the current BEV—the challenge of a support infrastructure, the issue of driving range, recent and ongoing battery developments, the issue of time to recharge. The chapter then considers the challenges of fuel supply—commercial infrastructure—domestic infrastructure—electricity infrastructure—BEV technology, and goes on to review dynamic electric charging trials, end-of-life recycling and recycling of BEV batteries and its challenges, in particular, those associated with lithium recycling and the chapter concludes with efforts to drive down the cost of BEV batteries.

Keywords Lithium · Cobalt · Thermal runaway · On-street parking · Charging · End-of-life recycling · Vertical integration

The Challenge of a Support Infrastructure

In Chap 2, we considered the different architecture requirements of different modes of transport. On the one hand, the quadruped requiring very little investment in infrastructure for the system to be operational—just some food and access to water. At the other extreme, the railway system requiring very considerable investment in cutting rail lines, laying rail tracks, boring tunnels, building bridges, building and staffing stations, and refuelling points, providing means to shunt and turn trains round, etc.—all of which had to be invested in, and put in place, before even the first journey can be provided.

And, somewhere in between, the arrival of the automobile—electric and gasoline powered, which needed a crude roadway system—the one already existing for horses and carts would do, and a fuel to power the vehicle.

And when the technologies had settled down the gasoline technology won out because the required infrastructure—a few tanks of gasoline scattered along the road network—was cheap and easier to provide; refuelling was quick and easy; and the cars were a lot cheaper. There was no similar support network for electric vehicles, and the lead-acid battery technology was simply not up to the task.

B. Williams, *Electric Shocks: Are Drivers Rejecting the Electric Car?*,
https://doi.org/10.1007/978-3-031-98820-2_4

As I concluded in the section 'The challenge of the internal combustion engine' way back in Chap. 2: 'Electric vehicles became popular for applications where they could be connected to mains supply of electricity (trains, trams and trolley buses) or their limited range did not pose major problems (Forklift trucks, milk floats and golf carts). But by the 1920s, the early heyday of electric cars had passed, and a decade later, the electric automobile industry had effectively disappeared'.

And as I noted in Chap. 3 'Dennis Minano, GM's vice president for Energy and Environment, questioned whether consumers desired electric vehicles. Robert James Eaton, chairman of Chrysler, also doubted the readiness of mass produced for electric cars, stating in 1994 that "if the law is there, we'll meet it ... at this point of time, nobody can forecast that we can make an electric car"'.

And I concluded in a later chapter about developments post-WWII.

> 'I think it is fair to conclude that for all the effort and investment made, in the period 1950—2000, the BEV was not fit for purpose...
>
> The major problems not overcome by all were range, recharging time, battery life and cost (although other design issues were also encountered by many)...
>
> The concept of the battery electric vehicle failed in 1950–2000 for the same reasons as its demise and fall in the first half of the century. Despite new pressing needs, the technical solutions were just not up to the job. Better solutions were available'.

What aspects of the game have changed or been addressed in C21 that may now make the BEV fit for purpose? What obstacles still remain? And what previously unknown obstacles have been unearthed/encountered that will hamper the success of the BEV?

The Issue of Driving Range

The biggest advances made in the period 2020–2024 have been in the range available from a single charge.

As of 2023–2024, several BEVs stand out for their impressive looking range capabilities.

GAC's Aion LX plus. LX Plus claims a range of up to 1008 km (626 miles).

US Lucid Air Dream Edition R claims a range of up to 837 km (520 miles).

Tesla Model S Long Range claims a range of up to 652 km (405 miles).

Mercedes-Benz EQS. 450+ claims a range of up to 563 km (350 miles).

BMW iX xDrive50 claims a range of up to 521 km (324 miles).

Ford Mustang Mach-E Extended Range claims a range of up to 502 km (312 miles).

Hyundai Ioniq 6 Long Range claims a range of up to 581 km (361 miles).

These selected examples are all 'top end' models likely to be chosen by those likely to make long car journeys (and are able to afford to do so). Their gasoline powered equivalents typically offer a range of 650–700 km (circa 400+ miles) on a tank full of gasoline, so half them appear to be equivalent in range. But I say 'appear' because all is not as it first seems.

The examples claim ranges of ‘up to xxx km’ because common and frequently encountered conditions adversely reduce achievable range. The range is a function of the chosen battery technology and size (measured in kWh) and the drawdown of power as the vehicle is in use/driven. The achievable range of a BEV is influenced by several key factors:

Driving Style and Speed: As with a gasoline powered vehicle, aggressive driving, such as rapid acceleration and high-speed driving, increases energy consumption. Prolonged motorway driving without much (regenerative) braking reduces range.

Temperature and Climate: Extreme cold or heat affects battery performance and efficiency. In cold weather, the battery’s chemical reactions slow down, and more energy is used to heat the cabin and battery, while very hot conditions can also decrease range due to the energy required to cool the cabin. Extreme cold can reduce battery efficiency by up to 40%.

Terrain: The type of terrain over which the vehicle is driven, such as hilly or flat affects range. Driving uphill requires more energy and reduces range, while downhill driving can extend range due to regenerative braking, which recaptures energy.

Vehicle Load: The weight of the passengers and cargo the vehicle is carrying. Heavier loads increase energy consumption, reducing range. This includes not just passengers and cargo but also accessories such as roof racks or trailers, which also affect the aerodynamic drag of the vehicle and further reduce range. The weight of the battery is also a significant adverse load.

Use of HVAC (Heating, Ventilation, and Air Conditioning): Using the HVAC system extensively, especially in extreme weather, can reduce range as these systems draw significant power from the battery.

Tyre Type and Pressure: The type of tyres (e.g. winter, summer, all-season) and their inflation level. Under-inflated tyres or tyres designed for specific conditions such as winter tyres can increase rolling resistance, reducing range. Properly inflated, low rolling resistance tyres maximise range.

Battery Age and Condition: Over time, a battery’s capacity diminishes due to degradation—its ability to hold a charge decreases, leading to a reduction in the vehicle’s achievable range.

Each of these factors can significantly influence the real-world range of a BEV, often making the actual range lower than the manufacturer’s advertised figures tested under ideal conditions. Of course, many of these factors affect gasoline powered and hybrid vehicles similarly, but range is not normally considered such a critical factor for those vehicles because of the ease of refuelling them.

Then there is a *difference in user expectation*. It is quick and easy to refill a gasoline powered vehicle, so it is a little inconvenient, but not serious, and can often be combined with a comfort break. It is a different matter recharging a BEV—first you have to find an available charging point, and then wait an age while it recharges. So the user expectation is to have enough useable range to make a long journey without the need to recharge en-route, requiring a significantly greater range than the gasolene powered equivalent.

The 80% Rule

To cap it all, while 'fast charging' has improved and the time to achieve an 80% recharge is significantly reduced, topping up the last 20% of a BEV battery is a process that can take longer and is less efficient than charging the first 80%. This is due to the way lithium-ion and similar BEV batteries are designed to protect their longevity and safety; therefore, as a BEV battery nears full capacity, the charging speed is slowed down significantly by the battery management system (BMS) to avoid overcharging and overheating, which could damage the battery cells and reduce their lifespan, and this reduces the speed and efficiency of topping up the last 20%. While controlling the heat of the battery, the energy efficiency decreases when charging the last 20% of the battery. This means more electricity is required to add the same amount of energy to the battery compared to the earlier stages of charging.

Further, charging a battery to 100% repeatedly can accelerate battery degradation over time. Most BEVs are optimised to operate best when charged between 20% and 80%, which helps extend battery life. Regularly charging only up to 80% is often recommended for daily use, leaving the full recharge for times when maximum range is necessary.

So a manufacturer may claim and test measure a range of say 600 km (370 miles), but not charging above 80% effectively reduces that to about 480 km, and the other degradation factors described above cut typically a further 25%, cutting the claimed 600 km range to an effective regularly achievable range to about 360 km (circa 220 miles) in most cases.

(As a guide the practical regularly achievable range is 55%–65% of the claimed range).

So, for the driver of the larger car intending to make longer journeys, BEV battery range has improved dramatically over recent years, but further improvement is still needed.

Development of BEV battery technology continues apace and is discussed further in this chapter, below. But in respect of the objectives of this work—identifying and solving problems with BEV acceptability, for larger cars/SUVs/vans the range problem is largely solved and because of the ongoing developments, summarized below, I think we can be confident will disappear as a problem by 2030.

Urban Environments

For use in urban environments, as a prime use vehicle or second vehicle, large or small, and indeed in most second vehicle situations since about 2015 range has not been a barrier to BEV ownership/use—indeed for many range was never an issue in the urban situation.

Smaller Cars

The second vehicle in a multi-car household is frequently used as an urban run-around by one or more members of the household and is often a 'small' car. But small cars are also used as prime vehicles by the less affluent, by older people, as 'first' cars, and increasingly by ecology-sensitive people. And while mid-sized and large sedans, SUVs and vans have space for larger batteries in order to achieve good range, the small, compact/subcompact BEV does not, and where the vehicle use is mainly urban runaround or relatively local family visits, as we have said above, range is largely not an issue. But for other users, range is important.

The space taken up by BEV batteries also reduces load-carrying space available. (The smaller the vehicle, the bigger that problem).

Studies have been made on the tolerance of different demographic groups to recharging a BEV during journeys and highlight several key factors. People's tolerance tends to vary based on factors such as age, lifestyle, and prior experience with BEVs.

Younger individuals, especially those who are tech-savvy, tend to be more tolerant of charging during trips, particularly if they perceive the charging stations as modern and accessible. Older drivers, used to rapid gasolene refuelling, tend to be less tolerant. The ecologically sensitive tend to be very tolerant.

Urban drivers, who have access to more frequent charging stations, may show higher tolerance for recharging during journeys compared to rural drivers. Rural drivers often express frustration with the limited availability of fast chargers.

The claimed range of current subcompact BEVs largely ranges between 230 and 400 km (140–250 miles), meaning practically achievable ranges of 130–200 km (80–125 miles), so for those straying beyond the urban environment and belonging to groups less tolerant to charging en-route, will have to rely on battery technology developments if range is not to be a deterrent to BEV ownership.

Recent and Ongoing Battery Developments

Several advanced battery technologies are being installed, or about to be installed in BEVs, or are in advanced stages of development:

Lithium Iron Phosphate (LFP): LFP batteries have become increasingly popular due to their safety, long cycle life, and lower cost. These batteries are widely used in Tesla's Model 3 and Model Y Standard Range models and are favoured by many Chinese manufacturers such as BYD. Offering high thermal stability, lower cost, and a long lifespan although they have a lower energy density compared to other lithium-ion batteries.

Nickel Cobalt Manganese (NCM): NCM batteries are known for their high energy density, making them suitable for longer range vehicles. The use of nickel increases energy density, while cobalt and manganese provide stability and longevity. Recent developments have focused on reducing cobalt content due to its cost and ethical concerns, leading to variants such as NCM 811 (8 parts nickel, 1 part cobalt, 1 part manganese).

Lithium Nickel Manganese Cobalt Oxide (NMC) 811: This specific chemistry variant within the NMC family has higher nickel content (80%) and lower cobalt content, which increases the energy density and reduces dependency on cobalt. It is used in premium EVs that require longer ranges, such as some versions of the Tesla Model S and Porsche Taycan.

Solid-State: Solid-state batteries are seen as the next big leap in battery technology, offering higher energy density, faster charging times, and enhanced safety due to the solid electrolyte that replaces the flammable liquid electrolyte in traditional lithium-ion batteries. Solid-state batteries are considered to be a significant advancement over traditional lithium-ion batteries, particularly in their resistance to thermal runaway.

Solid-state batteries are designed differently from traditional lithium-ion batteries in ways that make them less prone to thermal runaway. In solid-state batteries, the liquid electrolyte is replaced with a solid electrolyte, which can be made from materials such as ceramics, polymers, or glass. These solid electrolytes are non-flammable and are inherently more stable than liquid electrolytes. This reduces the risk of fires even when the battery experiences high temperatures or internal damage.

Solid-state batteries generally have higher thermal stability because the solid electrolyte is designed to not break down as easily as liquid electrolytes. This means they are more resistant to overheating and less likely to initiate the thermal runaway process.

Solid-state batteries are also less likely to experience internal short circuits, which are a major cause of thermal runaway in traditional batteries. The solid electrolyte acts as a barrier, preventing lithium dendrites (needle-like growths that can pierce the separator in lithium-ion batteries) from forming and causing a short circuit.

One challenge with solid-state batteries is the interface between the solid electrolyte and the electrodes. If this interface is not well-engineered, it can lead to mechanical stress and degradation, which might still cause overheating under certain conditions. However, this is less likely to lead to thermal runaway compared to liquid-based systems.

Solid-state batteries are currently more difficult and expensive to manufacture at scale. Creating defect-free solid electrolytes and ensuring good contact with electrodes is a complex process, which impacts their commercial viability.

Solid-state batteries can store more energy in the same volume, leading to lighter and smaller batteries with longer ranges for electric vehicles. The solid electrolyte in these batteries is more resistant to degradation, potentially allowing for more charge and discharge cycles and a longer overall lifespan.

Solid-state batteries may allow for faster charging times compared to traditional lithium-ion batteries, improving the user experience for BEVs.

While solid-state batteries are expected to begin appearing in consumer products and vehicles within the next decade, full commercial roll-out on a mass scale may take longer due to the challenges summarized above.

While still in the early stages of commercialisation, companies such as Toyota, MG, and QuantumScape are actively developing these batteries, with limited production expected to start by 2025. This could revolutionise the industry by significantly increasing the range and reducing charging times.

Cobalt-Free: In response to ethical and supply chain concerns around cobalt, manufacturers are developing cobalt-free or cobalt-reduced batteries. These typically involve variations of LFP or manganese-rich chemistries. Emerging in lower cost EVs and being tested in various high-volume models, particularly in China.

Sodium-Ion: Sodium-ion batteries (aka 'table-salt' batteries) are being explored as an alternative to lithium-ion batteries, with potential cost advantages due to the abundance of sodium. While they have lower energy density than lithium-ion, developments are being made to improve their performance. Companies such as CATL have announced plans to start mass production of sodium-ion batteries, expected to complement lithium-ion batteries in certain markets by 2025.

Silicon Anode: Replacing some of the graphite in the anode with silicon can significantly increase the energy density of lithium-ion batteries. Companies such as Tesla and start-ups such as Sila Nanotechnologies are working on this technology with the objective of increased energy density and potential for faster charging times.

Lithium-Sulphur: Lithium-sulphur batteries have the potential for very high energy density, but they face challenges with cycle life and stability. Ongoing research aims to overcome these issues, with some prototype vehicles being tested. If successful, lithium-sulphur could dramatically extend the range of EVs though they are still in the experimental stage.

These technologies, along with other more speculative developments, reflect the rapid pace of battery innovation in the BEV market, driven by the need for longer range, faster charging, reduced costs, less monopolistic dependencies, and improved safety. The adoption of these advanced batteries will continue to evolve as manufacturers refine these technologies and scale-up production.

The Issue of Time to Recharge

This is a problem area that won't go away.

Let us be clear. The overnight charging of a BEV at home is not foreseen as a problem for those with a garage or driveway attached to the house although it does require investment in the installation of charging equipment (about £1000–£1500 in the UK [US$ 1250–$1900]in some countries more).

Overnight gives plenty of time to recharge to 80%.

It is the time to recharge a BEV en-route—even assuming that you can find an available charging station. It is the time taken to recharge a BEV (to 80%) compared to the time it takes to refuel a gasoline powered vehicle that is the problem.

On average, including pay at the pump transaction, it takes less than 5 min to refuel a gasoline or diesel-powered vehicle. If you have to queue to get to the pump, then it can double, but is rarely more than 10 min in total.

Not so with refuelling a BEV.

The time it takes to recharge a BEV from 20% to 80% at a commercial charging station depends on the type of charger and the vehicle's battery capacity. As a general guide:

Fast Chargers (DC Chargers, typically 50 kW): Time: Around 40–60 min. - These chargers are common at service stations and can recharge most BEVs (such as a Nissan Leaf or Hyundai Kona) from 20% to 80% in less than 1 h.

Ultra-Fast Chargers (150 kW and above): - Time: Around 20–30 min. - With ultra-fast chargers, such as those found at certain motorway service areas or dedicated charging hubs, vehicles that can handle higher charging speeds (e.g. Tesla Model 3, Porsche Taycan) will recharge significantly faster using these ports.

Standard Public AC Chargers (7 kW to 22 kW): - Time: Anywhere from 2 to 4 h for a 7 kW charger. - These chargers are slower and more suited for longer stops or overnight charging but can still be used to top up a vehicle from 20% to 80%.

The actual time will vary based on the vehicle's battery size, the battery technology, the charger's power output, and the car's maximum charging rate.

Summary: This really is the first Achilles heel of BEVs. What time-pressed businessman wants to have to pause his day for half an hour to an hour just to refuel? Perhaps longer if the charging ports are busy when he arrives. No family, with kids on board, already bored by being cooped up in the car for hours, wants to experience these delays.

So they will do their best to avoid it…. Which makes it difficult to commercially invest in providing or upgrading these charging stations. (more on this below).

A combination of solving the range problem and improving battery technology to enable faster charging (both discussed above) seem to offer the only solution to alleviate this problem situation. And some of the developments summarised above show promise of truly rapid charging potential.

The Challenges of Fuel Supply (Commercial Infrastructure)

I have made several remarks in passing about the availability of commercial charging points en-route. And I point out here that this is the next Achilles Heel for the BEV driver. And to some extent an own-goal against governments pushing the transition to electric vehicles.

For, in most places in the world there is an inadequate number of commercial BEV charging points available.

This is the albeit somewhat crude but solid basis on which I make that assessment:

Taking the EU as a good example (because it comprises 27 diverse types of country).

There are some 280 million cars and vans (UNECE Categories M1/N1) in the EU.

These are supported by some 92,000 'gas stations' each with an average of say 8 filling points, i.e. say some 736,000 filling points. Assuming it takes 5 min per refuelling transaction, (12 transactions per pump/hour). That gives a total capacity of servicing over 8.8 million refuels per hour. 8.8 million divided by the vehicle population of 280 million means that the refuelling architecture could potentially service (refuel) just over 3% of the vehicle population in any hour.

In order to provide the same level of support of 3% of 280 million electric vehicles, even at just 30 min per recharge would require 4.2 million ultra-fast charging points. Currently (2024), the EU has just 96,000 ultra-fast chargers, indeed only 736,000 public charging points of all types.

It is true to say that all gasoline powered vehicles have to go to a service station in order to refuel every time, whereas most electric refuelling takes place at home, so there should be a big disparity in that need/requirement, and this is true, but it is largely also the case that *only* people who can recharge at home have BEVs at the moment and if all light vehicles have to be BEVs, that has to change.

You can with justification argue about what are the realistic figures all day long, but the fact that there is a hopelessly inadequate availability of public charging points in Europe to support a 100% BEV light vehicle population persists with any realistic numeric assessment.

Another way to look at it is that in the EU there is one public charging point per every 360 vehicles, one ultra-fast charger per 3000 vehicles.

By comparison, there are some 40.5 million cars and vans (UNECE Categories M1/N1) in the UK.

These are supported by some 8000 'gas stations' each with an average of say 8 filling points, i.e. say some 64,000 filling points. Again assuming it takes 5 min per refuelling transaction, (12 transactions per pump/hour). That gives a total capacity of servicing 0.77 million refuels per hour. 0.77 million divided by the vehicle population of 40.5 million means that the refuelling architecture could potentially service (refuel) about 2% of the vehicle population in any hour. (The average UK service station probably has 10–12 filling points, but I have kept with the same average size for comparison purposes.)

In order to provide the same level of support of 2% of 40.5 million electric vehicles, even at just 30 min per recharge would require 400,000 ultra-fast charging points. Currently, the UK has (2024) only 67,000 public charging points of all types.

In the UK, there is one public charging point per every 600 vehicles.

By comparison, China's investment in BEV charging infrastructure has been extensive, making it the world leader in this area. As of Q2 2024, China has deployed around 2.6 million public charging stations and an additional 2.6 million private home chargers. This gives the country a total of about 5.2 million charging points. To use the same comparison methodology.

There are some 300 million cars and vans (UNECE Categories M1/N1) in China.

These are supported by some 137,000 'gas stations' each with an average of say 8 filling points i.e. say some 1,096,000 filling points. Assuming it takes 5 min per refuelling transaction (12 transactions per pump/hour). That gives a total capacity of servicing 13 million refuels per hour. 13 million divided by the vehicle population of 300 million means that the refuelling architecture could potentially service (refuel) about 4% of the vehicle population in any hour. (The average Chinese service station probably has about six filling points because of the widespread rural communities but I have kept with the same average size for comparison purposes.)

In order to provide the same level of support of 4% of 300 million electric vehicles, even at just 30 min per recharge would require 12 million ultra-fast charging points, at 3%, 9 million. Currently, China has 2.6 million public charging points of all types and the government continues to subsidise investment in the roll-out of more.

In China, there is already one public charging point per every 115 vehicles.

(by comparison EU 1 every 360/1 ultra-fast every 3000; the UK 1 every 600)

The Chinese BEV charging infrastructure is largely owned and operated by a mix of state-owned enterprises, such as the State Grid Corporation of China, and private companies, with TELD being the largest private operator. The funding for this massive infrastructure comes from a combination of sources, including government subsidies, state-backed financial bonds, and private investment. For example, the Chinese government has facilitated this growth by providing substantial subsidies and financial incentives to both public and private sectors, enabling the rapid expansion of the network.

Additionally, the Chinese government has committed to further expanding the network, with plans to continue significant investments to support the growing number of electric vehicles on the road (BEV and PHEV). This investment is part of a broader strategy that aligns with China's long-term goals for reducing carbon emissions and leading the global transition to electric mobility.

As a result, some 26% of new vehicles sold in China in 2023 were BEVs (the EU 14.6%; the UK 16%).

It is important to note too, that China is not pressurising its auto-industry to convert to solely BEVs, but what it calls 'New Energy' vehicles (BEV and PHEV or fuel cell).

Conclusions: If a government wants to persuade its population to switch over to 'new energy' vehicles, of whatever type, the government has to invest in developing the necessary support infrastructure. It cannot rely on the commercial marketplace to adequately provide it because the impetus for change is not coming from marketplace demand. Therefore, it won't adequately happen without support.

This is particularly true in this case where the infrastructure architecture and time to recharge are seen by the motorist as a fundamental part of the problem, to be avoided wherever possible (which makes it more difficult to make a commercial financial return).

The Challenges of Fuel Supply (Domestic Infrastructure)

As if not already hobbled by Achilles Heels, the third, and perhaps the most difficult Achilles heel to obstruct the take up of driving (owning/leasing/long-term renting) a BEV, lies not in electric vehicle technology, but in the basic architecture of our cities, towns, villages, streets, and homes.

To use the UK as an example. The figures and percentage points vary from country to country, but the basic problem remains the same in all countries (except perhaps China which has only recently effected a more or less planned migration of much of its population to new cities/developments).

There are 40.5 million cars and light commercial vehicles in the UK. There are approximately 28 million homes in the UK. About 60% of homes in the UK have access to off-street parking, which includes both drives and garages. This translates to roughly 16.8 million homes with some form of private parking. It means that about 40% of homes, roughly 11.2 million homes, rely on on-street parking/communal car parks. That doesn't 1:1 equate with 40% of the 40.5 million light vehicle population because those with off-street parking are also more likely to own multiple vehicles.

Approximately 20% of UK homes are flats or apartments. This means there are around 5.6 million flats or apartments. Many of those apartments have shared off street parking but no access to a personal power supply. Many others have to rely on on-street parking. Conversely, many of the 5 m + light commercial vehicles are parked overnight on business premises.

It is difficult to obtain more precise figures, but probably true to say that in excess of 10 million vehicles—a quarter of the light vehicle park—do not have the means—nor the possibility of fitting the equipment—to recharge a BEV. The vast majority of that 25% of the light vehicle sector will therefore not buy/lease/rent a BEV.

What can be done to address this massive problem? (see typical example situation in Fig. 4.1).

Several initiatives have been developed in the UK and globally to address the challenge of recharging BEVs for those who rely on on-street parking. These initiatives aim to make charging more accessible in urban areas and for those without access to private drives or garages. But most of these initiatives face the same problems as any public charging point, discussed above, and most would require massive government investment, which, outside of China, is, as we have seen above, is noticeably lacking. Initiatives include:

Lamp Post Charging Stations: Retrofitting existing streetlights to serve as charging points for electric vehicles is perhaps one of the more innovative solutions. This makes use of existing electrical connections, but requires significant upfront investment by already cash-strapped local authorities, and complex payment systems which require, of course, each lamp post to become online. Companies such as Ubitricity have been deploying these systems in various cities across the UK.

Pop-up Chargers: Another proposed solution: Pop-up chargers are retractable charging points installed in pavements that can be raised when needed and lowered

Fig. 4.1 Typical North London street, with houses many divided into apartments and on-street parking::source CSi UK Ltd. Library

to remain flush with the ground when not in use. For example, the Urban Electric Network in Oxford has been trialling this innovative solution. This solution bears all the disadvantages described above for the lamp post solution, but in addition is even more expensive, indeed, very expensive to install, run, and maintain. The benefits of a pop-up road barrier are understood. The benefits of a pop-up charging point at a beauty spot are understood, but the benefits of a pop-up charger for on-street charging that is going to be in use much of the time, is harder to see.

Cable Channelling: Some councils are allowing residents to install cable channels or 'cable gulleys' across the pavement, enabling them to safely run a charging cable from their home to the street without obstructing pedestrians. For example, trials in cities including Oxford and London have allowed residents to recharge vehicles overnight by laying cables within these channels. But this solution presupposes that your vehicle is parked in front of your premises, which is not usually the case with on-street parking. And the insurance aspects (in respect of both the homeowner and the permitting council) also need careful consideration. May provide a solution for particular use cases (e.g. reserved disabled parking space in front of a house) but is unlikely to provide a general solution.

Community Charging Hubs: Dedicated charging stations in residential neighbourhoods, offering multiple high-powered charging points, located in residential areas as public charging points. Companies such as BP Pulse and GRIDSERVE are developing charging hubs across the UK, with some focused specifically on urban areas. But these suffer from the same investment need and inconvenience factors of any public charging point, and unless subsidised are another example resulting in an expensive price for the recharge, to be paid in most cases by the lowest income quartile householders.

Community charging hubs can work however, in apartment block or housing block communal car parks, particularly where there is a resident's association to help timetable access.

Expansion of public rapid charging stations located in key urban areas, shopping centres, and motorway services. These rapid chargers allow for quick top-ups for BEVs and reduce the dependency on home charging and may be adequate to keep low-mileage users topped up. But, as with any public charging point, requires upfront investment, and usually provide a significantly more expensive recharge than a home recharge.

Mobile Charging Solutions: Mobile charging vans providing on-demand charging services, coming to the user's location to recharge their BEV. For example, start-ups such as ZipCharge are working on portable chargers that allow BEV owners to top up their battery regardless of their parking situation. This results in another solution offering an expensive price for the recharge to be paid in most cases by the lowest income quartile householders, a valuable emergency service for an exhausted BEV battery, but unlikely to provide an affordable (or available) solution for the regular recharge.

Car Clubs with BEVs: Another solution being touted, and popular with city planners, are car-sharing schemes that provide access to BEVs. This, it is claimed, allows residents who do not have private parking to still use an electric vehicle without needing their own charging infrastructure, and is cost-effective as it removes the need to own a vehicle. For example, enterprise Car Club and Zipcar are expanding their fleets to include electric vehicles.

This option is not so much a solution for charging BEVs, as effecting a behavioural change. And these issues are dealt with towards the end of Chap. 1 in the section 'Social engineering, causes and consequences'.

And the business cases for companies such as ZIPCAR are proving difficult. Zipcar has struggled with profitability throughout its history. Despite being a pioneer in the car-sharing industry, it has yet to consistently turn a profit. When Zipcar went public in 2011, its stock initially surged, but it faced significant financial difficulties thereafter, with continued losses. Even after being acquired by Avis Budget Group in 2013, the company has not reported any sustained profitability. The acquisition was partly driven by Zipcar's financial struggles, with the hope that leveraging Avis's infrastructure would help improve its finances.

Zipcar generates revenue primarily through membership fees, hourly or daily car rentals, and additional fees for fuel and late returns. However, the high costs associated with fleet maintenance, insurance, and technology have kept it from achieving long-term financial success.

In summary, Zipcar profitability remains elusive, and after 13 years of trading, does not yet provide a sustainable business case.

Static Inductive Charging (Parked Vehicles): It has been claimed (and the claim supported by leading auto manufacturers) that the recharging process, involving physically connecting electric cables, is a significant deterrent to BEV take-up.

Static Inductive Charging enables BEVs to recharge while parked over an inductive pad. Companies such as WiTricity, Qualcomm Halo, and Plugless Power are developing systems for both home and public charging stations. Auto manufacturers such as BMW, Mercedes-Benz, and Hyundai have introduced models with inductive charging capabilities or have engaged in pilot programmes. Several cities, including Oslo and London, have begun installing wireless charging pads for electric taxis to reduce downtime and eliminate the need for plug-in chargers. But problems remain that need to be addressed:

Efficiency and Energy Loss: Inductive charging systems are typically less efficient than traditional plug-in systems. Energy losses occur due to the conversion of electricity into electromagnetic fields, which can result in 10%–20% less efficiency compared to wired charging. Ongoing research is improving coil alignment, power electronics, and wireless energy transfer methods to reduce energy loss.

Charging Speed: Inductive charging typically offers slower charging speeds compared to high-powered plug-in chargers. Current wireless systems are often limited to between 3.6 kW and 11 kW, whereas wired chargers can reach up to 350 kW. However, new developments aim to increase wireless charging speeds, with some experimental systems reaching higher power outputs (up to 200 kW).

Vehicle Alignment Sensitivity: Wireless charging systems require precise alignment between the vehicle and the charging pad to ensure optimal energy transfer. Misalignment can reduce charging efficiency or cause the system to fail entirely. Automakers are working on vehicle positioning technologies, including automatic guidance systems, to help drivers' park accurately over charging pads.

Compatibility and Standardisation: The lack of universal standards across different manufacturers and countries has led to compatibility issues, with some inductive systems only working with specific vehicles.

Durability and Weather Resistance: Inductive charging pads installed in outdoor environments face challenges related to weather, wear, and tear. Water, debris, or snow can interfere with the operation of charging pads. Engineers are developing more robust and weatherproof designs, including better enclosures and drainage systems to maintain functionality in adverse conditions.

Cost and Infrastructure: The cost of installing inductive charging infrastructure is significantly higher than that of traditional charging systems. Installing wireless charging pads in roads or parking spaces requires substantial investment and could therefore face slow adoption.

Expectations for the Future of Static Inductive Charging

As costs decrease and efficiency improves, inductive charging is expected to become more widespread, particularly for urban transportation (e.g. taxis, buses) and shared mobility services where ease of use is paramount. Ongoing research aims to boost the power output of wireless charging systems, making them more comparable to high-speed plug-in chargers. However, it is difficult to see any particular benefit it brings to on-street parking/charging. Indeed, the alignment requirements do not fit well with the random nature of most on-street parking.

Conclusion: A combination of these initiatives will address some of the BEV recharging issues for those without a garage or driveway, but will only address a minority, and with current technology, BEV recharging for most of the 40% of households that depend on on-street parking for 25% of the light vehicle population, remains an insurmountable problem.

The Challenges of Fuel Supply (Electricity Infrastructure)

The transition to BEVs poses challenges for the stability and management of the electricity grid (See Fig. 4.2). Widespread BEV adoption could lead to increased peak demand, necessitating upgrades to the grid and the development of smart charging solutions to manage load effectively.

Taking the UK as a typical example again—and most countries are in a similar position—if all cars in the UK were replaced by BEVs, the additional load on the national grid would be substantial. To understand this in detail, we need to consider several factors, including the number of vehicles, their energy consumption, and the current capacity of the UK's electricity grid.

According to recent statistics, there are approximately 35 million cars on the road in the UK. A typical BEV consumes about 15–20 kWh of electricity per

Fig. 4.2 Power pylons stretching across countryside: source CSi UK Ltd. Library

100 km. The average annual distance driven per car in the UK is around 12,000 km. This means that the annual energy consumption per BEV:

- At 15 kWh/100 km, this translates to 1800 kWh per year.
- At 20 kWh/100 km, this translates to 2400 kWh per year.

So, if all 35 million cars were replaced by BEVs the total additional energy demand would be.

- Lower Estimate: 35 million cars × 1800 kWh = 63 TWh per year.
- Upper Estimate: 35 million cars × 2400 kWh = 84 TWh per year.

To put that into perspective, the UK's total electricity consumption is around 300 TWh per year. Therefore, converting all cars to BEVs will increase electricity demand by approximately 21%–28%.

If a significant portion of BEVs were charging during peak hours (e.g. early evening), this could create substantial spikes in demand. The UK's current peak electricity demand is around 60 GW.

The UK's grid would require significant upgrades to handle this additional load, both in terms of generation capacity and distribution infrastructure. Investment in smart charging technologies, demand response, and increased renewable energy capacity would be essential to mitigate these impacts.

Bearing in mind that on average, it takes about 7–10 years to get an offshore wind farm from the initial project proposal to the point where it is generating electricity, and for a nuclear power station, the timeline is significantly longer, often taking 15 to 20 years from project proposal to electricity generation, I think that it is true to say that it is already impossible to make that transition by 2035.

The consequences (of 100% switchover to BEVs by 2040 [5 years after 2035 cut-off date for ICEV sales]) would likely be power cuts, electricity rationing and increases in electricity prices.

To put that on a global scale, the International Energy Agency estimate global electricity generation to be around 28,000 TWh.

If all passenger cars (M1) and light commercial vehicles (N1) became BEVs, and there are globally about 1.6 billion cars and light vans, each using an average of about 20 kWh per 100 km driven, and driving (including light commercial) an average of 15,000 km–18,000 km [9000–11,000 miles] per year = 3400 kWh per vehicle year. The total energy demand for 1.6 billion vehicles = 3400 kWh × 1.6 billion = 5440 TWh year. That represents an increase in generating capacity required of nearly 20%.

OK. Both the global and UK calculations are crude and could be out by 20%–25% either way (the vehicle population continues to expand rapidly and this is not factored into the calculations). But whatever the figure, it is significant!

With the timescales for new generation capacity described above, even if there was the funding, there would not be the time to implement. Governments around the world did not set their net-zero targets to 2050–2060 without cause.

And where is funding on this scale going to be found?

The Challenges of Fuel Supply (BEV Technology)

The three Achilles Heels so far identified in this work (time it takes to recharge; availability of recharging points; no ability to recharge at home) are intrinsic to the technical architecture of current BEV technical solution. If we can't change the obstacles, what can we change in the technical solution to avoid encountering the obstacles?

The most hopeful line of development is Dynamic Inductive Charging, but it is still at an early stage of development and faces significant issues.

Dynamic Inductive Charging (In-Motion Charging): I summarised the principles of static electric charging in the last section. The charging technology for dynamic inductive charging is similar in principle, but this technology allows vehicles to recharge while moving over specially equipped roads. The road contains embedded coils that transfer energy to the vehicle as it drives.

Sweden and Israel have launched projects to test dynamic charging for public transport and logistics vehicles as this technology is seen as particularly promising for electric buses, trucks, and other vehicles with high energy demands. But if adapted could also provide a more optimum solution for BEVs, overcoming all of the Achilles Heels described above.

Dynamic Inductive Charging works through electrified coils that are embedded beneath the road surface in designated lanes. These coils are energised with electrical current, creating an alternating electromagnetic field above the road. The road is divided into segments, with each segment activated only when a vehicle is passing over it to minimise energy waste and ensure safety.

Vehicles are fitted with receiver coils located on the underside of the vehicle. These coils pick up the electromagnetic field generated by the road coils and convert it back into electrical energy. The vehicle's energy management system directs this power to either recharge the battery or directly power the electric motor, depending on the vehicle's needs at the time.

When the vehicle drives over the charged road segments, the alternating electromagnetic field induces an electric current in the receiver coils of the vehicle. This wireless energy transfer is based on the same principles used in static inductive charging but adapted for dynamic conditions.

The efficiency of energy transfer depends on several factors, including the distance between the vehicle's receiver coils and the road's transmitter coils, vehicle speed, and coil alignment. The system is designed to deliver a controlled amount of power based on the vehicle's demand, ensuring safe operation for both the vehicle and the surrounding environment.

Implementing dynamic inductive charging would however involve significant infrastructure investment, and the question of how to finance and operate such systems is complex.

There are also questions regarding the infrastructure maintenance costs associated with such equipped roadways.

The potential models for financing, operating, and paying for in-motion charging are expensive and complex. The following have been suggested:

Subscription-Based Models: BEV users could subscribe to a dynamic charging service, paying a monthly or annual fee for access to inductive charging roads/lanes. Charging models similar to internet charging models could be applied.

Pay-Per-Use: BEV drivers might pay for charging dynamically as they drive, similar to toll roads. Sensors in the vehicle could track the amount of energy received, and users would be billed based on their consumption. This option may fit very well with a move to 'pay-per-mile' road taxation (some novel taxation solution will have to replace fuel duty and VAT on gasolene, and this benefit might make such tax more acceptable).

Integrated with Tolls: In countries with a wide network of tolled roads, charging fees could be integrated into existing toll road systems. As drivers pay tolls, part of the fee would cover the energy consumed from dynamic charging.

Sponsored Lanes: Companies, especially those in the green energy or automotive sectors, could sponsor dynamic charging lanes, with branding and advertisements used to offset costs.

Challenges to Suggested Payment Models

High Initial Infrastructure Costs: Building dynamic charging roads would be expensive. These costs include embedding coils, upgrading roadways, and installing energy management systems. Ongoing wear and tear maintenance costs are seen to be a significant problem as well. Without substantial public or private investment, the roll-out may be too limited to be effective. The financial model also needs to be scalable. If dynamic charging is only available on a few highways or urban areas, it may not attract widespread usage, limiting the usefulness of the solution.

Pricing Fairness: Determining how to fairly charge users for dynamic recharging will be challenging. High fees might deter use, while under-pricing could lead to financial losses for operators. Dynamic pricing models may need to adjust based on vehicle type, time of day, or overall electricity demand.

Dynamic Electric Charging Trials

Several major trials of dynamic inductive charging for BEVs are underway around the world. These trials aim to evaluate the feasibility, effectiveness, and scalability of dynamic charging technology. None are yet anywhere near commercialisation:

ElectReon in Sweden (Gotland Island): ElectReon, an Israeli company, is leading one of the most prominent dynamic inductive charging trials. This project involves the construction of a 1.65 km stretch of electric road between the city centre and airport on Gotland Island, Sweden. The road is designed to wirelessly recharge electric trucks and buses as they drive along it. The project is supported by the Swedish Transport Administration and includes partners such as Siemens and Volvo. The project aims to demonstrate the potential of dynamic charging for heavy-duty vehicles in cold climates, as well as to assess how well it integrates with existing road infrastructure. The trial has been underway since 2020, with successful demonstrations of in-motion charging for electric trucks.

ElectReon in Germany (Karlsruhe and Balingen): ElectReon is also leading a dynamic inductive charging trial in Germany. This project focuses on installing dynamic wireless charging infrastructure on a public road in Karlsruhe and Balingen, with the goal of testing the technology under real-world conditions. The project is supported by the German Federal Ministry of Transport and Digital Infrastructure and includes collaborations with public transport companies and automotive firms. The trial aims to evaluate the technology for use in public transportation and heavy-duty freight, with a focus on efficiency, durability, and cost-effectiveness. The German trial began in 2022, and tests are ongoing with promising initial results.

Stellantis and Arena del Futuro (Italy): Stellantis, in collaboration with several partners, is conducting a dynamic inductive charging trial called 'Arena del Futuro' in Italy. The project involves a 1 km circuit near Brescia that wirelessly charges electric vehicles while they drive. This trial involves partnerships with Fiat, Stellantis, A35 Brebemi (a motorway operator), and ABB. The goal is to demonstrate the feasibility of dynamic wireless charging for passenger cars and light commercial vehicles, with an eye towards improving BEV range and reducing the need for large onboard batteries. The system successfully demonstrated the ability to recharge vehicles in motion at highway speeds in 2022, and further trials are planned.

Indiana, USA (Purdue University): In the USA, Purdue University and the Indiana Department of Transportation (INDOT) are working on a dynamic wireless charging project. The first phase involves testing at Purdue's research facilities, with plans to install charging infrastructure on public roads for real-world testing. The project involves collaboration between Purdue University, INDOT, and German company Management. This project aims to develop and test a high-power dynamic charging system capable of charging BEVs as they travel at highway speeds, with a focus on freight and public transportation. Testing began in 2021, with plans for further expansion in the coming years.

ElectReon in Tel Aviv, Israel: Another significant trial by ElectReon is taking place in Tel Aviv. This project is focused on installing dynamic charging infrastructure on a section of road used by public buses. The project is supported by the Tel Aviv municipality and public transportation operators. The trial aims to demonstrate the viability of dynamic charging for public buses in urban environments, with the goal of reducing downtime for charging and minimising battery size. The trial has shown promising results since its launch in 2020, with buses successfully charging while on their routes.

Coventry, UK (Future Transport Zone): Coventry has been chosen as a test-bed for dynamic wireless charging under the UK's Future Transport Zones programme. This initiative will explore how dynamic charging can support electric taxis and buses. The project will focus on improving the efficiency of public transport fleets and reducing the need for extensive charging downtime. Trials are expected to begin in the near future, following the installation of the necessary infrastructure.

Elonroad, Lund, Sweden: EVolutionRoad is a project the Swedish Transport Administration assigned to build a pioneering test and demonstration site with Elonroad's unique electric road system in the city of Lund. This project is different from those above in that it is not contactless, but contact driven, ground level feed. The EVolution Road project stands as a pivotal initiative within the Swedish government's roadmap. The project develops, evaluates, and enhances the technology, aiming to document and disseminate crucial knowledge. Challenges, from friction and corrosion to charging powers, energy efficiency, electromagnetic emissions, driving speed, and varying weather conditions, are tested and the project navigates these complexities to pave the way for a sustainable and efficient electric road infrastructure.

This is the first electric road with a ground-level feeding system tested in an urban environment with a city bus and other vehicles. The project's consortium includes nine actors from the business sector, universities and the public sector: Innovation Skåne AB (project manager), Elonroad AB, Kraftringen Energi AB, Lund municipality/Future by lund, Lund University of Technology, LTH, Ramboll AB, Skånetrafiken AB, Solaris Sverige AB, National Road and Transport Research Institute, VTI. Duration 2019–2024. The Swedish Transport Administration financed SEK 83 million of the project's total budget of SEK 96 million. Other funding comes from partners that have joined forces in the Elväg Syd consortium.

The EVolution Road project boasts impressive outcomes from its intensive testing phase. The electromagnetic compatibility (EMC) performance has excelled, powering/recharging several BEV cars, a van and a bus, and proving the system's reliable operation even in challenging weather conditions such as heavy rainfall while maintaining minimal and harmless leakage currents. These critical findings are being shared with the Swedish Transport Administration to inform further development.

Currently, the project is delving deeper into various facets of the technology, including the nuances of operational procedures, usage trends, maintenance needs, and environmental impact. But while these trials are promising, challenges such as high infrastructure costs, efficiency optimisation, and ensuring the durability of road-embedded coils in various weather conditions, as does the charging payment model. Moreover, scaling dynamic charging across entire road networks requires substantial public and private investment. These trials are critical in refining the technology and demonstrating its economic and environmental benefits for wider adoption.

Summary

Dynamic inductive charging/ground feed charging have the potential to revolutionise how we power electric vehicles, particularly for fleets, public transport, and potentially BEVs. However, building the necessary infrastructure and developing sustainable payment models will require innovative approaches that involve cooperation between governments, private industry, and consumers. Much will depend on minimisation of the road infrastructure implementation cost per mile. The technology will only work for new generations of purpose designed vehicles. Global standardisation will be needed.

If dynamic charging becomes viable on a large scale, it could revolutionise how we think about electric vehicle range, reducing the need for large batteries and creating more efficient, always-on-the-go systems.

Future inductive charging systems may integrate with smart grids, enabling more flexible and efficient energy management, such as charging during off-peak times or using renewable energy sources when available.

Inductive charging holds significant promise for the future of electric mobility, but overcoming the many technical, economic, and standardisation challenges will be key to its viability and widespread adoption. Not a near-term solution.

End of Life Recycling

There are several 'end-of-life' issues associated with BEVs, but there are three major issues:

1. The normal recycling at the end of life of the vehicle.
2. The replacement of the BEV battery pack during the operating life of the vehicle.
3. The insurance/safety replacement of parts, including the battery pack following an accident.

Accident and insurance issues are dealt with separately in Chap. 6. But the principal end-of-life recycling problem lies in BEV battery technology, and principally with lithium and cobalt.

Lithium-ion batteries degrade over time, losing their capacity to hold a charge. This limits their lifespan and creates waste management challenges as more batteries reach their end of life. As lithium batteries age, their performance declines. This may cause their replacement during the working life of the vehicle, and of course disposal at the end of their life. Handling old or degraded batteries can pose increased risks of leakage, short circuits, or fires.

End-of-Life Processes for Recycling of BEV Batteries

The process and challenges involved in BEV battery recycling include:

Battery Collection and Sorting: At the end of a battery's life cycle, the battery must be properly collected and transported to a recycling facility. This can involve car manufacturers, dealerships, auto-recycling specialists and waste management services. The traditional auto scrap yard will in most cases be inadequate. Different types of batteries (e.g. lithium-ion, nickel-metal hydride) have to be sorted because the recycling processes vary depending on the battery chemistry.

Re-use in Energy Storage Systems: Before recycling, many BEV batteries, although no longer suitable for vehicle use, still have a significant portion of their capacity remaining. These batteries have the potential to be repurposed into energy storage applications, such as for renewable energy systems or grid stabilisation. But, on-the-ground, there is, as yet, very little evidence of this happening in practice.

Recycling Processes

- **Mechanical Processing:** This involves shredding the batteries and separating the materials (metals, plastics, etc.) through various methods such as sieving, magnetic separation, and chemical treatment.
- **Pyrometallurgical Process (Smelting):** This involves using high temperatures to melt down the battery components, which enable the extraction of valuable metals such as cobalt, nickel, and lithium.
- **Hydrometallurgical Process (Leaching):** This process uses chemicals to dissolve battery materials, followed by precipitation or solvent extraction to recover metals. This process is more energy-efficient and can recover a wider range of materials.

Challenges in BEV Battery Recycling

- **Economic Viability:** The cost of recycling often exceeds the value of the materials recovered, particularly for batteries with less valuable metals such as lithium-iron-phosphate.
- **Recycling Efficiency:** Not all materials can be fully recovered and losses occur during the recycling process.
- **Safety Concerns:** Handling and dismantling batteries can be hazardous due to the presence of toxic materials and the risk of thermal runaway. (see 'The challenges of working with the raw materials of BEV batteries' below).
- **Standardisation:** The variety of battery chemistries and designs complicates the recycling process, due to the lack of standardised battery designs and advanced

sorting technologies are required, making economic recycling more difficult to achieve.

- **- Legislation and Regulations:** In many countries, regulations require the recycling of a minimum percentage of the materials in batteries, and manufacturers are responsible for the end-of-life management of their products. However, the means to achieve this are, in many cases, as yet unachievable.
- **Recycling Challenges:** Although the recycling of lithium-ion batteries is theoretically possible, it is not yet widespread nor highly efficient. A significant amount of valuable materials can be lost during recycling, and the infrastructure for large-scale recycling is still underdeveloped.

Lithium Recycling Challenges

Recycling lithium-ion batteries is complex and an expensive process due to the different materials involved (such as lithium, cobalt, nickel, and others) and the difficulty in safely dismantling batteries without causing fires or releasing hazardous chemicals.

The process is, of course, to safely dismantle the battery and recover the valuable materials such as lithium, cobalt, and nickel. However, current recycling technologies are not always economically viable, especially for smaller batteries or those with lower concentrations of valuable metals. The cost of recycling lithium often exceeds the value of the recovered material, which discourages widespread recycling efforts. Additionally, the on-the-ground reality is that only a small percentage of lithium-ion batteries are currently recycled, leading to a growing waste management problem.

Recycling lithium-ion batteries is technically possible but, in practice, is not widely implemented. The infrastructure for recycling lithium and recovering valuable materials from batteries is still inadequate in many parts of the world. This leads to large amounts of lithium batteries being discarded rather than recycled.

Even with recycling, a portion of the materials in lithium-ion batteries may be lost or degraded during the process. This can reduce the overall effectiveness of recycling.

Improper disposal of lithium can lead to the release of toxic substances into the environment. When lithium-containing waste ends up in landfills, it can leach into soil and water, potentially contaminating ecosystems. High concentrations of lithium can be harmful to plants, animals, and aquatic life. When lithium batteries are incinerated in waste-to-energy plants, they can release harmful chemicals into the atmosphere, including dioxins and furans, which are dangerous to human health.

Recycling cobalt, particularly from products such as batteries, generates fine dust, which can be toxic if inhaled. Workers in recycling plants are at risk of developing lung diseases, including asthma and fibrosis, if proper safety measures are not in place. Direct exposure to cobalt during recycling can cause allergic reactions, skin irritation, and eye problems. Handling cobalt safely requires protective equipment, and inadequate precautions can lead to occupational health hazards.

Recycling cobalt is expensive, often outweighing the financial value of the recovered material, especially in small-scale operations. This creates a financial disincentive for companies to invest in cobalt recycling infrastructure, leading to higher rates of disposal rather than recycling. And even with advanced recycling technologies, the recovery rates of cobalt are often low, meaning that a significant portion of the cobalt in end-of-life products is lost during the recycling process. This loss reduces the overall efficiency of recycling and increases reliance on primary cobalt mining.

The extraction of cobalt through recycling involves energy-intensive processes, including pyrometallurgical and hydrometallurgical methods. These processes consume large amounts of energy, produce waste, and generate greenhouse gas emissions, reducing the overall environmental benefits of recycling.

Many regions lack the necessary infrastructure for efficient cobalt recycling, particularly for lithium-ion batteries. This results in low recycling rates and increased disposal in landfills. Building up the infrastructure for cobalt recycling requires significant investment in technology, training, and safety measures.

While the demand for cobalt in batteries has surged, recycling technology has not kept pace with this demand. Existing technologies are often not efficient enough to meet the recycling needs, leading to bottlenecks in cobalt recovery and an over-reliance on primary mining sources.

Driving Down the Cost of BEV Batteries

BEVs cost a third more than their ICEV equivalents, and this is discussed in several places in this work. The prime reason is the cost of the battery. So what are the auto manufacturers doing to improve the situation?

The cost per kilowatt-hour (kWh) for BEV batteries varies between manufacturers due to differences in technology, production methods, scale, and supply chain factors.

Tesla is one of the leaders in lowering battery costs, with OpenAI suggesting Tesla's battery costs are between $130 and $150 per kWh as of 2023. Tesla's close partnership with Panasonic and investments in Gigafactories have helped drive down costs.

Ford current costs are estimated by OpenAI to be between $130 and $150 per kWh for models such as the Mustang Mach-E and Ford is targeting $100 to $120 per kWh within a few years.

Currently, General Motors battery production costs are believed around $130 to $140 per kWh. GM is focusing on its Ultium battery technology, which OpenAI estimates could reach costs as low as $100 per kWh by mid-decade.

Volkswagen is targeting to reduce the cost of its batteries through its PowerCo initiative and aims to reach $100 per kWh by 2025–6. According to OpenAI, current costs are estimated at $140–$160 per kWh, slightly higher than Tesla and GM.

BYD, who are originally a battery manufacturer, now turned auto manufacturer benefits from producing their own 'blade' battery technology. Their battery costs are estimated by OpenAI to be around $120–$130 per kWh, one of the lowest in the sector.

All of the major OEMs are investing heavily in gigafactories or associating with partners to do so, with the ambition to get battery costs down to $100 per kWh well before 2030.

The race to reduce battery costs is critical as manufacturers aim to make BEVs more affordable and increase market penetration.

The Challenges of Working with the Raw Materials of BEV Batteries

The most common technology used in BEV batteries involves the use of lithium. The reaction of lithium with water is a notable chemical interaction, especially because of the potential safety and associated environmental hazards. When lithium (a highly reactive alkali metal) is exposed to water, it reacts to form lithium hydroxide (LiOH) and hydrogen gas (H_2).

Because the reaction between lithium and water is exothermic, i.e. the reaction generates significant amounts of heat, there is a very high risk that this heat can ignite the hydrogen gas produced, leading to explosion and/or fire. The release of hydrogen gas is dangerous because hydrogen is highly flammable. Therefore, when lithium-ion batteries are exposed to water, particularly when damaged, they create hazardous situations. This can lead to short-circuiting, fires, or even explosions due to the presence of both reactive lithium compounds, the flammable electrolyte in the battery and the hydrogen generated.

Lithium-ion battery fires generate significant heat, often reaching temperatures above 1000 °C (1,832 °F). This intense heat frequently leads to secondary fires and makes the fire difficult to extinguish. Lithium-ion fires frequently cause a condition known as 'thermal runaway', where an increase in temperature causes a chemical reaction that generates more heat, which further accelerates the reaction. This makes the fire self-sustaining and difficult to stop. In battery packs with multiple cells (like those in BEVs), thermal runaway in one cell spreads to neighbouring cells, causing a chain reaction where the entire pack of battery cells catch fire in an escalating cascading effect.

The rapid expansion of gases within the battery casing experienced during such a fire can cause a pressure build-up, leading to violent explosions.

Even after a lithium fire appears to be extinguished, lithium-ion batteries can reignite hours or even days later due to residual heat or internal chemical reactions. This makes monitoring and additional fire suppression efforts necessary after the initial fire has been put out.

Lithium-ion battery fires also release a range of toxic gases, including hydrogen fluoride (HF), carbon monoxide (CO), and other harmful organic vapours. The vapours produced, especially hydrogen fluoride, can be highly corrosive and cause damage to equipment, surfaces, and the lungs if inhaled. These gases pose significant health risks to firefighters and bystanders and may require specialised protective equipment for those responding to the fire.

When lithium-ion batteries catch fire in confined spaces (e.g. inside a BEV), the release of heat and toxic gases poses significant risks to occupants and responders. The confined environment can accelerate the spread of the fire and concentrate toxic emissions, making evacuation and firefighting more difficult.

Lithium and lithium-ion battery fires are unique due to their high heat, potential for re-ignition, toxic gas emissions, and in some cases, violent explosions. Effective fire suppression and management often require specialised techniques and equipment to ensure safety and mitigate risks. This makes lithium battery fires particularly challenging for firefighters and emergency responders.

In order to prevent accidental contact with moisture in the air or water, lithium metal needs to be stored in non-reactive environments, typically in mineral oil or other non-water-based environments. To make the situation even more volatile, lithium is also highly reactive with oxygen. When exposed to air, it rapidly oxidises, forming a layer of lithium oxide on its surface. In high humidity conditions, this can accelerate and lead to hazardous reactions. Lithium in its raw form needs to be handled in dry, controlled environments with proper ventilation. Therefore, lithium metal must be stored under an inert atmosphere (such as in oil, argon, or another non-reactive gas) to prevent it from reacting with moisture or oxygen in the air.

(And, incidentally is one of the reasons that BEV batteries are so heavy because they have to be very securely waterproof encased to prevent any chance of contact with water encountered by a vehicle in very wet or flood conditions, or the ingress of air).

If lithium comes into contact with skin, it can cause chemical burns, especially in its reactive form. Direct exposure to lithium hydroxide (formed when lithium reacts with water) can be particularly harmful and corrosive to skin and eyes.

Handling lithium therefore requires stringent safety protocols. Workers dealing with lithium need to use protective gear, including gloves and eye protection, to prevent accidental contact.

The improper disposal of lithium can result in environmental contamination. When lithium is released into the soil or water, it can have toxic effects on ecosystems. High concentrations of lithium in water can be harmful to plants and aquatic organisms.

If they are not properly recycled, lithium batteries contribute to the growing e-waste problem because they can leak hazardous chemicals, including lithium compounds, into the environment, posing long-term risks to soil and groundwater.

BEV batteries also often include cobalt.

One problem with handling Cobalt is that Cobalt is toxic when inhaled, ingested, or when it comes into direct contact with the skin. Exposure to cobalt dust or fumes can cause lung diseases, such as ‘hard metal lung disease’, a form of interstitial lung disease. Chronic exposure to cobalt can also lead to respiratory problems, asthma, and fibrosis. Direct skin contact with cobalt can cause allergic dermatitis, leading to rashes, redness, and skin irritation.

The International Agency for Research on Cancer (IARC) has classified cobalt as a possible human carcinogen (Group 2B), particularly in its soluble forms. Long-term exposure to cobalt compounds, especially through inhalation, is linked to an increased risk of cancer, especially in the respiratory system. Cobalt is often processed in powder form, which creates a significant inhalation hazard. Workers involved in mining, refining, and processing cobalt are at risk of inhaling fine cobalt particles, leading to respiratory illnesses. Proper ventilation and dust control measures are critical to prevent exposure.

Handling cobalt safely requires the use of personal protective equipment, such as gloves, respirators, and protective clothing. Inadequate protective measures can lead to health problems for workers, especially in industries such as battery manufacturing.

The production of Cobalt's production is highly concentrated in a few countries, with the Democratic Republic of Congo (DRC) accounting for more than 60% of global production. This makes the cobalt supply chain vulnerable to political instability, labour strikes, and other disruptions that can lead to supply shortages and price volatility.

The extraction and refining of cobalt have significant environmental consequences. Cobalt mining, particularly in regions such as the DRC, often leads to deforestation, soil degradation, and water pollution. Toxic byproducts from cobalt processing can contaminate local water supplies, harming both ecosystems and human communities.

The refining process of cobalt produces hazardous waste, including heavy metals and toxic chemicals, which require careful management and disposal to prevent environmental contamination. Improper handling of cobalt waste can lead to soil and water pollution, posing long-term risks to the environment.

Recycling cobalt from BEV batteries is complex and therefore expensive. Recovering cobalt from lithium-ion batteries, for example, requires sophisticated recycling techniques that are not yet widely implemented.

Components of lithium-ion batteries, such as electrolytes, are potentially toxic and pose *environmental risks*.

So, lithium-ion batteries, and its variants, are the most widely used technical solutions used in BEV batteries, and present several significant challenges and problems:

The first major challenge is *Thermal Runaway*: as stated above, lithium-ion batteries can experience a phenomenon called thermal runaway, where heat generated inside the battery leads to a rapid, uncontrollable increase in temperature. This can result in fires or explosions, particularly when the battery is damaged or improperly handled. To exacerbate this further, the electrolyte used in lithium-ion batteries is often also flammable, increasing the risk of fire if the battery is punctured or short-circuited.

The second challenge is the *Loss of Capacity over Time.* Lithium-ion batteries degrade over time, losing capacity after repeated charge and discharge cycles. This degradation limits the lifespan of batteries, especially in high-demand applications such as BEVs, where performance and range are critical.

The next challenge is *Calendar Ageing*. Even when not in use, lithium-ion batteries lose capacity due to calendar ageing, which is influenced by temperature and storage conditions.

Another challenge is *Temperature Sensitivity.* Lithium-ion batteries perform poorly in extreme temperatures. Cold temperatures reduce their capacity and efficiency, while high temperatures accelerate degradation and increase safety risks.

A further concern is *Geopolitical Risk.* A significant portion of the global supply of lithium, cobalt, and other battery materials comes from politically unstable regions, which can lead to supply chain disruptions and increased costs.

Handling of both lithium and cobalt involves both toxic and handling risks, and significant *risk of fire and explosion.*

The End-of-Life Management of lithium batteries is therefore difficult and expensive. The variety of lithium-ion battery designs and alternate chemistries makes recycling more difficult. The economic viability of recycling lithium batteries remains a challenge. In many cases, the cost of recycling is higher than the value of the materials recovered, discouraging widespread recycling efforts.

The Carbon Footprint not only of manufacturing, but also of recycling, lithium–ion and similar batteries is energy-intensive and results in significant carbon emissions, particularly in countries where production relies heavily on fossil fuels. Combined with the fact that in most countries 49%–70% of the electricity used during their working lifetime is not 'green', has raised concerns about the overall carbon footprint of BEVs (more on this subject in 'Zero tailpipe emissions v lifetime emissions' Chap. 5 below).

Chapter 5
New Energy Vehicles

Abstract Chapter 5 discusses 'New Energy Vehicles' and compares the strategies of western countries with that of China. The chapter addresses the issue of zero tailpipe emissions vs lifetime emissions. The chapter then examines PHEVs and in particular the Chinese strategy regarding PHEVs and BEVs. Automotive solar panels, battery swap vehicles, fuel cell electric vehicles (FCEVs), liquid hydrogen ICE (HICEV), and butanol are assessed.

Keywords New Energy Vehicles (NEV) · Battery Electric Vehicles (BEV) · Plug-in Hybrid Electric Vehicles (PHEV) · Fuel Cell Electric Vehicles (FCEV) · Emissions · Lithium · Cobalt

Different Strategies

While the focus of vehicle electrification in the west has been the drive to BEVs, the focus in China has been a drive to 'New Energy Vehicles'. The Chinese official definition of a New Energy Vehicle (NEV) includes vehicles that utilise alternative energy sources instead of traditional fossil fuels. According to the Ministry of Industry and Information Technology (MIIT) of China, NEVs are categorised into the following types:

1. **Battery Electric Vehicles (BEVs)**: These are fully electric vehicles that operate using electricity stored in batteries. BEVs do not have an internal combustion engine and produce zero tailpipe emissions.
2. **Plug-in Hybrid Electric Vehicles (PHEVs)**: These vehicles have both an internal combustion engine and an electric motor. The battery can be recharged by plugging it into an external power source, allowing the vehicle to operate in all-electric mode for a certain range before switching to hybrid mode, or the vehicle is driven by electric power and the ICE is used to recharge the battery while driving to extend its operating range.

B. Williams, *Electric Shocks: Are Drivers Rejecting the Electric Car?*,
https://doi.org/10.1007/978-3-031-98820-2_5

3. **Fuel Cell Electric Vehicles (FCEVs):** FCEVs are powered by hydrogen fuel cells, which generate electricity to power the electric motor. They emit only water vapour as a byproduct and are considered zero-emission vehicles.

So far this work has dealt principally with BEVs, its focus subject. It is time look at the alternative technologies to reduce hydrocarbon emissions.

Zero Tailpipe Emissions vs Lifetime Emissions

Auto manufacturers like to claim that their BEVs are 'zero emissions' vehicles. Indeed, the Nissan Leaf claims it on its rear door.

But advertising control authorities are increasingly stamping down on such sweeping claims as misleading (see Fig. 5.1). For example, the UK's Advertising Standards Authority recently ruled (against BMW and MG) that 'any "zero emissions" claim for electric cars that didn't make it explicitly clear it related to emissions while driving, was likely to mislead'. When it came to the BMW advert, the ruling stated that 'without material information to make clear to consumers what the claim was based on, and clarification that it was restricted to emissions only when driving, we concluded "Zero Emissions Cars" was likely to mislead'.

It is clear that an internal combustion engine creates tailpipe emissions, and a BEV does not. But when comparing zero tailpipe emissions to lifetime emissions, it is important to understand that these two concepts address different aspects of a vehicle's environmental impact:

Zero Tailpipe Emissions During Driving: Vehicles with zero tailpipe emissions (BEVs and FCEVs) produce no direct emissions of pollutants or greenhouse gases (such as CO_2, NOx, or particulate matter) from their exhaust. But note, they still are the source of other emissions during driving.

Lifetime emissions: Aside from tailpipe emissions, cars, during their manufacture and use over their working lifetime, produce several other types of emissions that contribute to environmental degradation and air pollution. These emissions can occur during the manufacturing process, while driving (non-tailpipe related), and

Fig. 5.1 Production BEV claiming to be 'Zero Emission'source CSi UK Ltd. Library

even during maintenance and disposal. But although they are emissions, and potentially harmful, they are not all carbon generating emissions. The main sources of these emissions being:

Non-Tailpipe Emissions during Driving

Tyre Wear Emissions: Microplastics, rubber, and Particulate Matter (PM2.5): As tyres wear down, they release tiny particles of rubber, microplastics, and particulate matter into the environment. These particles are a significant source of non-exhaust emissions and contribute to air and water pollution. Tyre wear is an important consideration when comparing BEVs and ICEVs, both in terms of maintenance costs, and microplastic and particulate emissions into the air and especially into rivers. More frequent renewal of tyres, a factor that affects BEVs, is, in itself, also a contributor to carbon emissions because the process of making tyres is energy intensive.

Although both types of vehicles experience tyre wear and its emissions, several factors specific to BEVs influence how BEV tyres wear out more rapidly (several studies indicate in the region of 20% more rapidly). The greatest of these being the weight of the vehicle. BEVs are significantly heavier than their ICE counterparts due to the weight of the battery pack. For example, a Tesla Model 3 weighs around 1600–1900 kg, whereas a similar-sized ICE vehicle may weigh 1300–1400 kg. The 25%–30% extra weight increases the load on the tyres, causing them to wear out more quickly than on a lighter ICE vehicle. Heavier vehicles typically generate more friction and stress on the tyre tread, leading to accelerated wear, especially on the front tyres.

BEVs are also known for their rapid torque delivery, which means they can accelerate rapidly from a standstill. This is due to the electric motor's ability to deliver maximum torque almost immediately, unlike ICE cars that gradually build torque as the engine revolutions accelerate. This often gives the BEV stunning performance, but the instant torque can lead to more aggressive wear on the tyres, especially during rapid acceleration, where the tyres are digging in to drag the weight of the vehicle forward. This wear is much more severe in a BEV than a lightweight high-performance sports car because of the heaviness of a BEV. The tyres need to handle more stress as they grip the road, which can cause faster deterioration of the tread compared to an ICE vehicle with more gradual torque delivery. Ergo, more emissions.

BEVs, FCEVs, and hybrids use regenerative braking to recover energy during deceleration, which reduces the need for conventional braking. When the driver lifts their foot off the accelerator, the electric motor acts as a generator, slowing the car down and recharging the battery. Regenerative braking reduces the reliance on traditional brakes, which can lower wear on the brake system. However, since regenerative braking still requires the tyres to grip the road for deceleration, it can put additional strain on the tyres. The effect on tyre wear is complex and varies depending on how aggressively regenerative braking is used.

Internal combustion engine vehicles generally experience less aggressive tyre wear and cause less microplastic and particulate emissions because they are lighter and do not have instant torque.

Brake Wear Emissions: When brakes are used, friction between the brake pads and the rotors generate small particles of dust, which are released into the air as fine particulate matter (PM2.5 and PM10). This dust often contains metals such as copper, iron, and zinc, which can be harmful to both human health and the environment. This is a source of emissions for both BEVs and ICE vehicles, but because the BEV is so much heavier it uses the brakes harder, and wears brakes down more quickly.

Road Surface Wear: The interaction between tyres and the road surface causes wear, by both BEVs and ICE vehicles, releasing particles from the road itself (particulate matter). These particles contribute to air pollution, especially in urban areas with high traffic density. There is more or less a direct relationship between vehicle weight and road surface wear and tear, so the BEV, which is much heavier, does more damage. Indeed, many relate the 'pothole' crises experienced in many countries to the growth of the BEV population, but the trend to (heavier) SUVs and economizing on road maintenance must take their share of the blame as well.

Energy Source Emissions (Indirect Emissions): The energy source emissions from the production of petrol and diesel are well known and one of the reasons (in addition to tailpipe emissions) for the move to BEVs. But BEV energy source emissions depend on the energy source used for generating the energy used to recharge the battery. If that electricity comes from fossil fuels (such as coal, oil, or natural gas), there are emissions that are created associated with generating that electricity. These emissions contribute to the overall carbon footprint of the vehicle. At the time of writing, only around 30% of utility-scale generating capacity in the USA is generated from renewable carbon-free sources. In Europe, around 42% of the EU's electricity is generated from renewable sources. In China, it is 35% and in UK in 2023, 51% of electricity was generated from zero-carbon sources.

This means that, at the moment, the dream of BEVs being emission free regarding their consumption of fuel, is just a pipedream, because around the world 49–70% of electricity is generated from CO_2 emitting hydrocarbons. That will change over time of course, but it raises into question the rush to enforce BEVs ahead of the adoption of 'green' energy production. (More on this subject below).

Manufacturing Emissions

The production of a car, be it a BEV or ICE vehicle, involves significant energy use and emissions from mining and refining raw materials such as steel, aluminium, lithium (for batteries), and other metals. The extraction of these materials causes CO_2 emissions, habitat destruction, and water pollution. The following paragraphs are focused on the differences in manufacturing emissions between BEVs and ICE vehicles during the processes of manufacturing (and are not a summary of all emissions generated during manufacturing [e.g. emissions are created when rust-proofing and painting a vehicle, but there is no significant difference between doing this for a BEV or ICE vehicle]).

An ICEV engine consumes more energy to build than a BEV motor because of the energy used to make the engine block. But the biggest difference between BEV and ICEV manufacturing emissions lie in the manufacture of the BEV battery. In comparison to the simple manufacture of a fuel tank, the production of a BEV battery pack involves significant emissions, primarily due to the energy-intensive processes required to extract, refine, and assemble the raw materials, as well as the manufacturing process itself. The key causes being:

Raw Material Extraction and Refining

- **Lithium:** A critical component in lithium-ion batteries, the metal lithium is extracted through mining and brine processes, particularly in regions such as Australia (hard rock mining) and South America (lithium brine). These extraction processes can lead to significant environmental impact, including habitat destruction, high water usage, and energy consumption, contributing to CO_2 emissions.
- **Cobalt:** Cobalt is mined, predominantly in the Democratic Republic of Congo. Cobalt extraction often involves energy-intensive processes, and in some cases, artisanal mining practices (small-scale, often informal mining operations carried out by individuals, families, or small groups using basic tools and techniques. These operations are typically labour-intensive, and miners rely on rudimentary methods rather than advanced machinery or large-scale infrastructure) that increase emissions and, due to the dangers associated with handling cobalt (see above), also raise human rights concerns.
- **Nickel:** Used in the battery's cathode, nickel mining, and refining are energy-intensive and can lead to considerable greenhouse gas emissions.
- **Graphite:** Graphite is used for the anode in many lithium-ion batteries. Natural graphite extraction and synthetic graphite production both involve significant energy use and contribute to emissions.

Manufacturing Process

The battery manufacturing process itself is highly energy-intensive, and the majority of emissions come from:

- **Material Processing:** Purifying and processing lithium, cobalt, nickel, and other materials for battery-grade purity is energy-intensive.

- **Cell Manufacturing:** Assembling battery cells involves further emissions, especially in energy-demanding processes such as drying the electrodes, assembling the cells, and applying solvents.

Carbon Footprint Estimates

The size of the battery pack (measured in kilowatt-hours [kWh]) significantly impacts the emissions. Larger packs, such as those used in long-range electric vehicles, will generally result in more emissions than smaller packs. Typical ranges of emissions are measured in kilogrammes of CO_2 equivalent (kg CO_2-equivalent) per kWh of battery capacity. Estimates suggest that the production of lithium-ion battery packs can result in emissions between 56 and 494 kg CO_2-equivalent per kWh of battery capacity. For example, a 60 kWh battery pack could result in 3360–29,640 kg CO_2-equivalent emissions, depending on the energy mix used during production.

As electricity grids decarbonize and renewable energy becomes more prevalent, the lifetime emissions of BEVs and FCEVs will continue to decrease, making them more environmentally favourable. But while BEVs and FCEVs have zero tailpipe emissions, their lifetime emissions are significant, and depend on factors such as manufacturing processes and the source of electricity or hydrogen used.

Plug-in Hybrid Electric Vehicles (PHEV)

Plug-in Hybrid Electric Vehicle (PHEV) technologies represent a hybrid of two major automotive power sources: internal combustion engines (ICE) and electric motors powered by mains electricity rechargeable batteries (BEV technology). PHEVs have historically been designed to offer flexibility, increased efficiency, and lower emissions by allowing drivers to use electric power for short trips and gasoline or diesel for longer journeys.

The hybrid element comprises one of three powertrain architectures:

- **Parallel Hybrid System:** This is the most common PHEV (and ICE hybrid) configuration, where both the internal combustion engine and the electric motor can independently drive the vehicle. The system switches between or combines the two power sources based on driving conditions and battery charge levels.
- **Series Hybrid System:** In this system, the internal combustion engine does not directly power the wheels. Instead, it acts as a generator to recharge the battery, while the electric motor drives the vehicle. This design is less common in PHEVs but is historically used in certain models, particularly using a small gasoline tank and small engine to top up the battery as a range extender in smaller cars with limited space for the BEV battery.

- However, a new generation of series hybrid vehicles, often larger vehicles, are now providing significant range advantage, overcoming range anxiety, and providing flexibility for those without the ability to charge overnight at home. Some current (2024) examples are given below.
- **Series-Parallel Hybrid System:** Some PHEVs combine aspects of both series and parallel systems, allowing for greater flexibility in how the car uses its internal combustion engine and electric motor.

PHEV Battery Capacity and Range: Typical historical battery capacity in PHEVs in smaller vehicles has been smaller than in full BEVs, because of space, weight, and cost issues. These batteries can store enough electricity to provide pure electric driving for distances typically ranging from 20 to 60 km (12–37 miles) or more, depending on the model.

This smaller capacity is designed to strike a balance between electric driving range and maintaining a reasonable cost and weight for the vehicle. But the new generation of serial PHEVs (EREVs) are often larger vehicles focused on providing considerable range, and therefore have batteries of a capacity closer to that of a full BEV (but still smaller to keep the cost below that of a full BEV).

Battery Management Systems (BMS): As with any hybrid powered vehicle, a PHEV operates a sophisticated battery management system which ensures the longevity, safety, and efficiency of the battery. The BMS monitors and optimizes charging, manages heat distribution, and prevents overcharging or deep discharging.

Regenerative Braking: As with any hybrid powered vehicle, a PHEV operates regenerative braking systems, which capture the kinetic energy generated during braking and convert it into electricity that is fed back into the battery. When the driver applies the brakes or decelerates, the electric motor functions as a generator, recovering energy that would otherwise be lost as heat. This process helps extend the vehicle's electric range and improves overall energy efficiency.

Engine Start/Stop Technology: As with any hybrid powered vehicle, parallel PHEVs use ICE engine start/stop technology, which shuts off the internal combustion engine when the vehicle comes to a stop (e.g. at traffic lights) and automatically restarts it when the ICE is needed.

Thermal Management: Maintaining an optimal temperature is crucial for battery performance and longevity. PHEVs typically use liquid or air-cooled systems to manage the battery's temperature during charging and operation. In cold climates, PHEVs may also include battery heating systems to ensure efficient operation.

Electric-Only Driving Mode: Like many hybrid vehicles, parallel PHEVs typically feature an electric-only (EV) driving mode, allowing the car to operate purely on battery power for a certain distance, depending on the battery's charge. But whereas on a hybrid vehicle where the electric only range is limited to a few miles at most (often less) a PHEV EV range is typically 20–60 km (12-37 m), and in some recent cases quite a lot more. This mode is designed for urban driving and short commutes, reducing fuel consumption and emissions. (more on this below).

Charging: Unlike standard hybrids, PHEVs have an external charging port that allows the battery to be recharged from a standard electrical outlet or dedicated BEV charging station. Charging times vary based on the battery size and charger capacity, but as the PHEV battery size is in most cases limited, most PHEVs on the market can be fully charged in 2–4 h using a Level 2 charger (240 V). and 30 min or less using a rapid charging point, though the new generation long-range serial PHEV EREVs require longer.

Smart Charging: Like many more recent BEVs, some PHEVs are equipped with smart charging capabilities, allowing drivers to schedule charging during off-peak hours or prioritise charging when renewable energy is available. This can help reduce charging costs and the vehicle's environmental impact.

Advanced Engine Technology: Many PHEVs are powered by smaller, efficient internal combustion engines since rapid acceleration is efficiently achieved by the electric motor and low-speed driving is in any event more efficient using the electric motor(s) so long as the battery has power available.

Drive Modes: Many PHEVs offer selectable drive modes, allowing the driver to choose between different performance settings such as EV-only mode, Hybrid mode, Sport mode, or Charge mode (where the internal combustion engine charges the battery while driving).

China and PHEVs

Unlike most western government initiatives, PHEVs are core to China's 'New Energy Vehicles' strategy to lower carbon emissions.

China is a major player in the production of Plug-in Hybrid Electric Vehicles (PHEVs), and several Chinese automakers offer models with impressive electric ranges. Some recent examples are given below:

Chery Fulwin T10

- Electric Range: Up to 130 miles.

The Fulwin T10 is a compact SUV that features a 1.5-litre four-cylinder turbo engine, capable of 151 bhp and generating 162 lb. ft. torque, coupled with Chery's Super Hybrid Dual Mode PHEV system. That system is powered by a 34.5 kWh battery and comes in two guises: a less powerful 369 bhp front-wheel drive version (that can achieve 130 miles range) and a four-wheel drive version, producing 601 bhp and making 723 lb. ft. torque.

Li Auto L8

- Electric Range: Up to 110 miles (177 km).

The Li Auto L8 is a midsize luxury SUV with a significant electric-only range. It is designed for long-distance driving with, the makers say, a focus on comfort and advanced features. It achieves its impressive electric range through a combination of advanced hybrid technology and large battery capacity.

Importantly, the vehicle utilises an 'Extended-Range Electric Vehicle' (EREV) powertrain, which incorporates both an electric motor and a small internal combustion engine (ICE) that acts as a generator, where the gasoline engine does not drive the wheels directly but acts as a generator to recharge the battery when it depletes (series hybrid technology). This allows the vehicle to maintain electric operation for longer distances by supplementing the battery with additional power generated by the engine. The system prioritises electric driving whenever possible and engages the generator engine only when necessary, which preserves the battery's charge and extends the range. This approach extends the range compared to traditional plug-in hybrid systems.

Li Auto also make the **Li Auto L9.**

- Electric Range: Up to 131 miles (210 km).

The Li Auto L9 is a luxury full-size SUV with an extended electric range using the same technologies Extended Range Electric Vehicle (EREV).

The Li Auto L8 and L9 are manufactured by 'Li Auto', a Chinese automotive company that specialises in the production of EREVs, which was founded in 2015 to produce premium smart electric vehicles with extended-range (EREV) technology. Li Auto is one of the leading new energy vehicle (NEV) manufacturers in China, and the Li Auto L8 and L9 are part of their lineup of luxury extended-range electric SUVs. In 2023, Li Auto sold a total of 376,030 vehicles, reflecting an impressive 182.2% increase compared to the previous year (source CarNewsChina.com).

BYD Song Plus DM-i

- Electric Range: Up to 71 miles (114 km).

The BYD Song Plus DM-i is a midsize SUV that uses BYD's DM-i hybrid system, providing a competitive electric range for its segment. It is popular in China for its practicality and efficiency.

It achieves its electric-only range through a combination of what it calls a 'DM-i (Dual Mode Intelligent) powertrain' and a large well-optimised battery. The DM-i powertrain is a PHEV system that prioritises electric driving, with the electric motor taking precedence for most driving scenarios, especially at low speeds and in city driving. The gasoline engine primarily acts as a generator, charging the battery when needed or assisting the electric motor during high-demand situations such as highway driving or acceleration. (series-parallel hybrid technology).

Chery Tiggo 8 Plus PHEV

- Electric Range: Up to 62 miles (100 km).

The Chery Tiggo 8 Plus PHEV is a midsize SUV offering a decent electric range, making it a practical choice for families seeking to reduce their fuel consumption while maintaining flexibility for longer trips.

It achieves its range thanks to a 19.26 kWh battery and the efficiency of good design.

BYD Tang DM

- Electric Range: Up to 62 miles (100 km).

The BYD Tang DM is a large, luxury SUV that combines high performance with a solid electric range. It is one of BYD's flagship PHEVs, known for its spacious interior and advanced technology.

It achieves its range thanks to a 21.5 kWh battery and the efficiency of good design.

The role of the PHEV in the reduction of carbon emissions is discussed further in Chaps. 7 and 8.

China and BEVs

The last section looked at China and PHEVs. It is worth taking a little time to consider China and BEVs, and to compare its structure to its equivalents in Europe and USA.

While the structure of the auto manufacturing industry in USA and Europe is well organized and well established over the past century, it has been organized to manufacture and sell ICEVs, and it is having to adapt to BEV production not because of market pressure for BEVs, but because of government pressure to force BEV use, particularly in Europe. The sector has had to not only absorb the cost of developing and installing different technologies and manufacturing methods, but invest heavily in marketing to persuade the consumer to buy the more expensive BEV products. To a large extent, it is burdened by its organized past, rather than benefitting from it, and this at a time of other rapid technology advancements and commercial pressures leading to structural rationalization and reorganisation.

By comparison, the structure, organisation, and integration of the Chinese BEV sector reflects the country's rapid industrialisation and heavy government involvement, and in particular the tenth revision of the Chinese Government's strategic '86/3 Program', that was issued in 2001 and included the development of 'New Energy Vehicles' -BEVs, hybrids and fuel cell vehicles (as a National strategic goal, and supported by heavy government investment in order to achieve it), and the associated strategic investment of the 86/3 Program has resulted in China becoming the world's largest market for, and producer of, electric vehicles.

As a result, whereas the Chinese market for, and manufacture of, ICEVs, evolved initially through joint ventures with globally dominant auto manufacturers, Chinese OEMs dominate the BEV and PHEV market (see above).

It means that whereas the USA and European auto manufacturers came to implement BEV technology largely in the period post 2015, Chinese industry already had 10–15 years of heavily subsidised investment and experience, from a clean sheet, to call on.

Battery production is a crucial part of the Chinese BEV sector, with Chinese companies leading not only the domestic market but now the global market as well. It includes companies such as CATL (Contemporary Amperex Technology Co. Limited), the world's largest producer of lithium-ion batteries, supplying batteries to automakers worldwide; BYD. a significant battery manufacturer turned auto manufacturer (It is vertically integrated, producing both BEVs and batteries [see above]); and CALB, Gotion High-Tech.

China's BEV sector also has an extensive and integrated supply chain, including manufacturers of electric motors, power electronics, control systems, providers of charging stations, connectors, and related hardware.

Massive government subsidies, through the government's tenth revision of the Chinese Government's strategic '86/3 Program', for both manufacturers and consumers, have been key to the growth of the sector. Although subsidies are gradually being phased out, they have already established the market in China.

Significant state funding has been allocated and continues to be allocated to R&D for advanced battery technologies, automated driving, and smart vehicle technologies, assisting Chinese companies to take and maintain its lead in the technological innovation of BEVs.

Since 2001, China has reportedly invested over US$ 60 billion (source OpenAI) directly into the 'New Energy Vehicles' sector through a combination of subsidies, grants, and incentives for manufacturers, infrastructure projects (charging stations, etc.), and R&D initiatives.

Starting from a clean sheet and heavy investment (and as with Tesla), many Chinese BEV companies have, as strategic business strategy, become highly vertically integrated, controlling multiple aspects of their value chain. As reported extensively in multiple sections of this work, BYD is a prime example of vertical integration, as it produces its own batteries, electric motors, and vehicles, reducing dependency on external suppliers.

The strategy of vertical integration, it is claimed, allows companies to control production costs; reduce reliance on global supply chains (and associated tariffs imposed by foreign governments) and provide supply chain stability; innovate more rapidly (integration of R&D, manufacturing, and sales allows for faster development of new technologies and product launches).

Vertical Integration is a different philosophy to the 'Just-in-Time (JIT) Global Supply Chain' strategy that has served the auto manufacturing industry so well over the past half century.

Both vertical integration and just-in-time global supply chains offer distinct benefits and drawbacks when it comes to managing production, costs, and efficiency.

Vertical integration requires a significant initial investment to acquire or build different parts of the supply chain. This includes the setting up and operation of manufacturing capabilities, frequently requiring diverse skill sets. Operating multiple parts of the supply chain, from raw materials to distribution, requires managing complex processes across different business areas, which can lead to inefficiencies.

Further, owning parts of the supply chain locks a company into certain fixed costs (e.g. factories, equipment, labour), reduces flexibility to adapt to demand fluctuations or market changes.

When a company integrates vertically, it may lose focus on its core competencies (e.g. product design, marketing) by also managing upstream (suppliers) and downstream (distribution) functions. Specialised suppliers and logistics providers may have economies of scale and expertise that allow them to be more efficient than an integrated company managing those processes in-house. With vertical integration, companies may struggle to adapt quickly to market shifts or changes in consumer demand because they are tied to their internal production capacity and suppliers, unlike JIT models where companies can switch suppliers easily, and if internal production cannot meet demand, the company may face shortages, while companies using external suppliers can seek out additional capacity through already established channels.

Owning multiple stages of the supply chain increases risk exposure. For instance, a disruption at any stage (e.g. raw material shortages, factory failures) could impact the entire operation, as the company lacks external fallback options and in economic downturns, vertically integrated companies may struggle to reduce costs quickly because they are tied to their own supply chain investments, unlike companies with a JIT supply chain that can scale down or shift suppliers. But vertical integration does lessen the risk of/exposure to quotas and tariffs on components imposed by foreign governments.

JIT systems on the other hand are designed to keep inventory levels low by ordering parts or products only when they are needed. This reduces warehousing costs, inventory holding costs, and the risk of inventory obsolescence. This means that instead of tying up capital in large inventories, companies can allocate resources to other areas, such as innovation, marketing, or growth initiatives. This improves cash flow and reduces the need for expensive storage facilities.

JIT systems enable companies to source from a global network of suppliers, allowing them to choose the most cost-effective or specialised suppliers based on current needs. This flexibility ensures companies can adapt quickly to changes in demand without being constrained by large stockpiles of materials or components.

JIT encourages lean manufacturing, where inefficiencies are minimised, and waste is reduced. By receiving parts and components only when needed, companies can streamline production processes, avoid overproduction, and reduce storage costs. With access to a global supply base, companies can benefit from lower-cost suppliers in different regions, particularly in low-cost manufacturing countries, which further drives down production costs. Since JIT systems do not rely on large inventories, companies can more easily adapt to new technologies, processes, or trends. They are not left with obsolete inventory, allowing them to innovate faster and stay competitive.

While just-in-time global supply chain systems provide the advantages of lower operational costs, flexibility, and lean production, it comes with risks such as supply chain disruptions, reliance on global suppliers, and exposure to quotas and tariffs imposed by foreign governments for political reasons. Vertical integration offers the benefits of greater control, higher margins, and secure supply chains.

The choice between these models often depends on the specific needs of a business. Companies in industries where supply chain security is paramount and have rapidly evolving technology, which means that there is a lack of an established market supply, which is typical at the current time for the fast evolving BEV product, favours vertical integration in order to control production costs, provide supply chain stability and innovate more rapidly.

When the market stabilizes and matures over time, with its focus becoming more on cost efficiency and agility and market price pressures, rather than product development and evolution, akin to traditional auto manufacturing, it might be better adapted to a JIT global supply chain model, and have to write off many of the costs of its investment in vertical integration.

Through the government's tenth revision of the Chinese Government's strategic '86/3 Program', the Chinese government has also accelerated the development of an extensive charging network, which has proven to be important to public acceptance of BEVs.

Europe has become the second-largest market for BEVs. The European Union (EU) is pushing for widespread electrification, and countries such as Germany, Norway, the UK, and France are at the forefront of implementation. However, the market is more fragmented due to different national government policies, infrastructure readiness, and consumer preferences in each country.

The EU has set stringent CO_2 emissions targets, forcing automakers to produce more BEVs to avoid penalties. Many European countries also offer subsidies for BEV purchases, lower taxes, and other incentives. The EU's 'Green Deal' and 'Fit for 55' initiatives aim for carbon neutrality of its Member States by 2050, with a focus on transitioning to electric mobility. Countries such as Norway have extremely high BEV adoption rates due to favourable policies (e.g. no purchase taxes for BEVs, road toll exemptions), while Germany and France offer significant subsidies for BEV buyers.

Europe is investing heavily in local battery production to reduce dependency on China. Gigafactories by companies such as Northvolt, Volkswagen, and ACC (a joint venture between TotalEnergies and Stellantis) are under construction or already in operation.

The USA is behind both China and Europe in BEV adoption. In 2023, BEVs accounted for around 7–8% of new car sales. The USA has a mix of federal tax credits for BEVs and various state-level incentives. States like California lead in BEV adoption due to strict emissions regulations and incentives, but many other states lag behind.

The Biden administration announced large-scale investment in EV charging infrastructure, aiming for a nationwide network of 500,000 chargers by 2030 as part of the Infrastructure Investment and Jobs Act. The Trump administration is cutting back.

The USA is way behind China in battery production capacity but is trying to catch up with domestic initiatives. But the USA is heavily reliant on imported battery materials, leading to concerns about supply chain security and the need for investment in domestic raw material production and processing.

China's 20 years of coordinated and funded strategy, starting with a clean sheet, contrasts with Europe's fragmented market and fragmented and far weaker funding and initiatives to a well-established sector that was organized to produce something else, and the USA's slower but emerging transition, create distinct regional approaches to the BEV sector, with China clearly winning out at this stage.

Automotive Solar Panels

The Netherlands Organization for Applied Scientific Research (TNO) has conducted research that supports the potential of BEVs incorporating photovoltaic (PV) modules on their bodywork or roof to significantly extend their driving range through solar energy alone. According to their findings, BEVs equipped with integrated solar panels have the potential to achieve more than 10,000 km per year of driving purely from solar power under optimal conditions. TNO's research shows that solar cells integrated into a vehicle's roof, hood (bonnet), or bodywork can generate substantial amounts of electricity, depending on the geographical location and the amount of sunlight available.

The TNO research highlights ongoing improvements in the efficiency of solar cells, as well as in the design and integration of PV modules into car bodies. These advancements are making it more feasible for solar energy to meaningfully contribute to a BEV's overall range.

The distance a solar-powered car can travel using solar energy depends heavily on its geographic location. While in sunnier areas such as southern Europe the figure could exceed 10,000 km per year, even in northern Europe, where sunlight is less intense, solar panels could still contribute a significant amount of energy.

In the opinion of TNO, the integration of PV modules would reduce the need for frequent charging from the grid, especially for short daily commutes, which could provide a major convenience for drivers. In some cases, solar energy could cover the entire energy needs for daily commuting, making the vehicle almost self-sufficient. The combination of solar energy and electric drivetrains could lead to a new class of highly efficient vehicles that rely less on external charging infrastructure, contributing to a more sustainable and decentralised energy ecosystem.

The TNO's findings emphasise the transformative potential of solar integration in BEVs, particularly as solar technology becomes more efficient and cost-effective.

(Source: ChatGPT and TNO)

Do Solar Panels recharge When the Vehicle is Off and Locked?

Yes, in most cases, the solar panels continue to work even when the vehicle is off and locked, as long as the car is exposed to sunlight/bright daylight. The solar cells are independent of the vehicle's ignition system and can recharge the battery or power systems when parked. The extent of charging depends on the system design and the amount of sunlight available.

This offers a particular advantage for our problem area of on-street parked vehicles because they are, by definition, almost all exposed to sunlight when parked during the day.

A few car models currently incorporate solar panels to enhance energy efficiency and range. These panels are typically integrated into the roof or other parts of the vehicle, providing a supplementary energy source for either the vehicle's electrical systems or directly contributing to charging the battery.

Hyundai Ioniq 5: The Hyundai Ioniq 5 offers a solar roof option, designed to supplement the electric range of the vehicle. The solar panels collect sunlight and convert it into electricity, which is used to recharge the car's 12 V battery and assist the high-voltage battery. Hyundai claims that under optimal conditions, the solar roof can add up to 2000 km (1242 miles) of range per year. Charging When Off and Locked: the solar roof continues to recharge the battery even when the vehicle is off and locked, as long as there is sufficient sunlight.

Toyota Prius Prime: Earlier models of the Toyota Prius Prime, specifically in Japan and Europe, had an optional solar roof. The solar panel primarily powered the car's auxiliary systems, such as air conditioning and interior electronics. The solar panels on the roof generate electricity to help power small electrical systems and reduce battery drain, enhancing the car's overall efficiency. In some versions, the solar energy could be used to slightly recharge the hybrid battery. Charging When Off and Locked: In these earlier Prius Prime models, the solar panels could continue to work when the vehicle was parked to power ventilation systems or trickle-charge the battery. That Toyota discontinued this as an option says something about the systems popularity or cost efficiency. But note that this was an early implementation, and solar panel technology has improved greatly in recent years.

Mercedes-Benz VISION EQXX: The VISION EQXX BEV working concept car model by Mercedes-Benz has 117 solar cells on its roof. Mercedes-Benz has unveiled the latest model of its electric fleet, Vision EQXX, whose ultra-thin roof panels feed the battery system to add up to 25 km (15.53 miles) of extra range on long-distance journeys on a single day and under ideal conditions, (source: Mercedes-Benz.) The energy generated by the solar cells is stored in a lithium-iron-phosphate battery and is used to power the climate blower, the lights, the infotainment system, and other ancillaries. In the future, the carmaker plans to use Mercedes-Benz solar power to recharge the high-voltage system, too.

The company claims that it is the most efficient car it has ever built with energy consumption of less than 10 kWh per 100 km and a range of more than 1000 km on a single charge. The car is equipped with a battery pack with almost 100 kWh of energy. The battery is, however, 50% smaller and weighs 30% less than the already benchmark pack in other EQS models in order to fit into a compact car.

The vehicle was developed in cooperation with the Fraunhofer Institute for Solar Energy Systems ISE.

Hyundai Sonata Hybrid: Hyundai Motor's solar roof charging system makes its debut on the New Sonata Hybrid. The system supports the vehicle's electric power source, improving fuel efficiency and reducing CO_2 emissions.

In the coming years, Hyundai will roll out the technology to other vehicles across its range.

The solar roof system includes a structure of silicon solar panels that are mounted on the car's roof. The system is composed of a solar panel and a controller. Electricity is produced when solar energy activates the solar panel's surface, which converts this energy by using photons of light from the sun. This creates the electron-hole pairs in silicon cells, which generate solar electricity. The electricity from this process is converted to the standard voltage by the controller, then stored in the battery.

Being able to recharge even while driving, the solar roof system can recharge 30–60% of the battery per day. With 6 h of daily charging, it is expected to increase drivers' travel distance by an extra 1300 km annually. (source Hyundai Motors).

Toyota BZX4: An optional solar panel roof is available for the Toyota (beyond zero) bZ4X in selected markets. This allows renewable solar energy to be captured and used for charging the battery or operating auxiliary systems, via a solar ECU located under the rear seats. In a sunny climate, Toyota estimates sufficient power to travel 1800 km could be obtained, or 11.7 km per day (equivalent to 140 smartphone charges).

When the car is parked with the ignition turned off, power from the solar panel is used to recharge the main battery, ensuring a sufficient distance to empty and allowing power to be supplied for other purposes, such as electrical appliances. When the car is being driven, with the ignition on, the solar power is supplied to car's auxiliary systems, reducing the draw on the main battery.

Developments in Solar Power Technologies

Recent developments in solar power technologies have led to significant advancements in efficiency, cost-effectiveness, and versatility. These innovations are transforming the potential of solar energy, making it more competitive with traditional energy sources and expanding its application across various sectors. Some of the key developments in solar power technologies that may be suitable for BEVs are summarised below:

Perovskite Solar Cells: Perovskite solar cells are a breakthrough in solar technology due to their high efficiency, low production cost, and flexibility. Perovskites are a class of materials that have shown tremendous potential for improving the efficiency of solar cells. Perovskite solar cells have achieved efficiencies of over 25% in laboratory settings, rivalling traditional silicon-based cells. They can be produced using simple manufacturing techniques such as printing or roll-to-roll processes. The main challenges with perovskite solar cells have been stability and durability. However, recent research has made significant strides in improving their long-term performance and resistance to environmental factors such as moisture and heat.

Bifacial Solar Panels: Bifacial solar panels can capture sunlight from both sides of the panel, increasing the amount of energy generated. These panels are installed on reflective surfaces such as white or light-coloured car roofs and hoods, allowing them to also absorb reflected light, thus increasing their efficiency. Bifacial panels can increase energy output by 10–30% compared to traditional single-sided panels.

Tandem Solar Cells: Tandem solar cells combine two or more layers of photovoltaic materials, each optimized to capture different parts of the solar spectrum. This allows them to convert more sunlight into electricity than single-junction cells. By stacking different materials (such as silicon and perovskite), tandem solar cells can achieve efficiencies exceeding 30%, significantly higher than conventional silicon solar cells. The complexity of manufacturing tandem cells and ensuring compatibility between different materials remains a hurdle, but progress is being made towards commercial viability.

Solar Skin Designs: Solar skin is a new aesthetic technology that allows solar panels to be customised with designs that match the appearance of the environment (such as car body panels). Its technology aims to make solar panels more visually appealing and blend in with the surroundings. While the primary focus is aesthetic integration, manufacturers are working to ensure that solar skins do not significantly reduce the energy efficiency of the panels.

Transparent Solar Panels: Transparent solar panels, or solar glass, can be used in sunroofs to generate electricity while allowing light to pass through. While transparent solar panels are less efficient than traditional opaque panels, ongoing improvements are increasing their power output.

Quantum Dot Solar Panels: Quantum dot solar panels are a cutting-edge technology that uses tiny semiconductor particles only a few nanometres in size to capture and convert sunlight into electricity. Quantum dots have unique optical and electronic properties due to their size, which makes them highly efficient at absorbing and emitting light. This technology holds promise for creating more efficient, flexible, and affordable solar panels compared to traditional silicon-based solar cells. One of the major advantages of quantum dots is their ability to absorb light across a broader spectrum of sunlight. They can be tuned to absorb different wavelengths, including infrared light, which is typically missed by conventional silicon solar cells.

Quantum dot solar cells have the potential to achieve higher efficiency than traditional solar panels. They can be designed to utilise more of the solar spectrum, including lower-energy photons like infrared light, potentially increasing the overall energy conversion efficiency. Quantum dots can be manufactured using relatively inexpensive chemical processes and can be applied as thin films on various surfaces, including flexible substrates. This reduces material costs and allows for cheaper, large-scale production. Unlike rigid silicon panels, quantum dot solar panels can be flexible, opening up new possibilities for integration into a wide variety of surfaces, such as windscreens and curved surfaces.

One of the challenges with quantum dot solar cells is ensuring long-term stability. Quantum dots can degrade under prolonged exposure to light and air, which can affect performance over time. And while the theoretical potential of quantum dot solar cells is very high, researchers are still working on improving their efficiency in practical applications to compete with or surpass traditional silicon-based panels. Quantum solar dots are still a way from commercialisation, but the technology shows promise for BEV solar panels in the longer term.

Conclusion

Solar power technologies are advancing rapidly. The improvements in perovskite solar cells, bifacial panels, quantum solar dots, tandem cells, and innovations in solar integration into everyday materials such as windows and roofs are expanding the possibilities for solar energy. Quantum dot solar technology represents a promising future for solar energy, with the potential to revolutionise how solar panels are used to provide power to BEVs, but are still a long way from commercialisation.

With continued investment and research, solar power is poised to become a significant energy source for BEVs, although is unlikely to become the main source of power without some, as yet unforeseen, development.

Battery Swap Vehicles

Actors

The Chinese ‘New Energy Vehicles’ initiative includes ‘battery swapping’ as an alternative to recharging. Aulton, BAIC, Geely, and NIO responded. So far, NIO is the automaker that has made most progress utilising battery swapping technology for BEVs (both in China and Europe). NIO sold about 160,000 BEVs in 2023.

In China, NIO has a growing network of over 2400 swap stations.

NIO also operates in Norway, Germany, Netherlands, Sweden, and Denmark.

Norway was the first European country where NIO entered. NIO opened its first battery swap station in Norway in 2022, and by mid-2023, it had four swap stations. NIO plans to further expand this network across the country.

NIO expanded to Germany in 2022 and has built several swap stations there, including near major cities like Berlin and Munich. The number is expected to grow as the company pushes further into Europe.

NIO launched its BEVs in the Netherlands in 2022. As of mid-2024, NIO is actively building battery swap stations, with plans to establish at least 20 stations by 2025.

Sales began in Sweden in 2022. One battery swap station is operational with plans to expand.

Denmark is also part of NIO's sales pitch. In Denmark, NIO is also gradually establishing battery swap stations, but there are fewer than 5 as of 2023, with expansion plans.

NIO faces several challenges in expanding its battery swap infrastructure:

- High cost of infrastructure: Building a battery swap station costs significantly more than traditional charging infrastructure, which limits rapid global expansion.
- Regulatory differences: Different standards for EV batteries in various countries make it challenging to standardise battery swapping.

BAIC offers a range of electric vehicles designed with battery swapping capabilities, including the EU300 and EU5 models. These vehicles are tailored for battery swaps, particularly in urban fleet and ride-share operations where minimising downtime is crucial. BAIC is partnering with Aulton battery swap stations.

Geely has also entered the battery swap BEV market, recognising the growing demand for quicker and more convenient charging alternatives. One of Geely's vehicles, Geometry C has been modified for battery swapping. This model is aimed at fleet operators who benefit from minimising downtime during recharging. Geely also has other vehicles under its brand Lynk & Co, which have integrated battery swapping technology.

Geely has collaborated with Aulton and battery manufacturer CATL to advance its battery swapping plans. Geely has also formed partnerships with other automakers and suppliers to standardise battery swapping technology.

Aulton is a specialist battery swap provider. It now operates hundreds of battery swap stations, primarily in China, including cities like Shanghai and Chongqing. By the end of 2023, Aulton was projected to operate 300 stations in Shanghai and over 200 in Chongqing. The company has ambitious plans to expand further, with a target of building 10,000 swap stations by 2025, designed to serve over ten million BEVs. Aulton claims that its battery swap technology can complete a battery swap in about 20 s, and its stations are geared towards servicing commercial and fleet vehicles.

Despite its rapid growth, Aulton's focus remains on China, where its partnership with Sinopec, a state-owned oil and gas provider, is helping it scale up by adding battery swap facilities at gas stations across the country. Challenges for Aulton (and other battery swap companies) include ensuring compatibility across different EV models and overcoming the higher infrastructure costs compared to regular charging stations.

China's government is heavily backing the battery swap model, leading to a significant number of swap stations, particularly for taxis and commercial fleets. The country has over 5000 battery swap stations, with projections for rapid growth in the coming years.

Advantages of Battery Swap over Recharging

- **Time-saving:** Battery swaps can be completed in as little as 3–5 min, providing a much faster alternative to recharging, especially compared to current fast-charging times.
- **Longevity:** Battery swapping can help increase the longevity of BEVs by allowing older batteries to be replaced with new ones, thus reducing the concerns over battery degradation.
- **Financial Flexibility:** The batteries are frequently provided on a lease basis, removing the battery cost from the purchase price of the vehicle. Considerably reducing the upfront cost of a BEV (then making the comparison between the battery lease costs and gasoline usage costs).
- **Convenience:** The ability to swap batteries rather than wait for recharging offers convenience, particularly in urban environments with high vehicle turnover.

Disadvantages of Battery Swap

- **High Infrastructure Costs:** Building a network of swap stations requires significant investment in infrastructure. The stations are more complex than regular charging stations due to the need for advanced mechanical systems to handle battery swapping and storage.
- **High Operating Costs:** As well as requiring one or more mechanics to make the swap, significant working capital is required to maintain a stock of ready-charged batteries. These costs are on top of the electricity costs of recharging the batteries.
- **Vehicle Compatibility:** Only works with BEVs that are specifically designed for battery swapping.
- **Ownership Model:** Battery swapping typically separates the ownership of the battery from the car itself, which may create complications in markets where consumers are used to owning both. Leasing batteries can raise concerns about ongoing long-term costs and with reliability issues.
- **Battery Standardisation:** Although China is working towards standardising battery designs, global standardisation remains a challenge. Different manufacturers produce batteries of varying sizes, shapes, and chemistries, which complicates the development of a universal swapping system.

- **Stifles Improvement of Battery Technology:** Because the need for a 'standard' battery locks the technology into that available today.
- **Limited Adoption Outside China:** While battery swapping is growing rapidly in China, it has struggled to gain traction in other regions like Europe and North America. Many automakers outside of China prefer to focus on fast-charging technologies and larger battery capacities, making battery swapping less appealing.

Conclusion

Battery swapping has shown strong potential in China, where government support and heavy infrastructure investment have helped it gain momentum, but it is still a small proportion of BEVs in China. However, its success globally will depend on overcoming significant hurdles, including high infrastructure costs, high operating costs, the need for standardisation, and wider consumer adoption. Despite these challenges, companies like NIO are proving that battery swapping can be a viable solution for reducing charging times and improving the overall BEV experience in specific markets.

Fuel Cell Electric Vehicles (FCEVs)

As with BEVs, the FCEV is driven by an electric motor, or, in the case of larger vehicles, two electric motors, one for front wheels and one for rear wheels. But the batteries that power the motors, or the electric motors themselves, or a combination of the two, are charged by hydrogen fuel cells, a device to convert the chemical energy from a 'fuel' into electricity through a chemical reaction with oxygen or another oxidising agent.

Fuel cells are different from batteries in that, as with internal combustion engines, they require a constant source of fuel and oxygen to run, but they can produce electricity continually for as long as these inputs are supplied.

The exciting thing about FCEVs is that the source of the compressed 'green' hydrogen they use as fuel is water, which is plentiful, and the only emission after extracting the hydrogen from water (H_2O) is oxygen, and the only tailpipe emission from the fuel cell process in the vehicle using the hydrogen (explained below) is water/water vapour. If the hydrogen is extracted using 'green' electricity, you have an ideal carbon emissions free vehicle that can be refuelled quickly at service stations in the same manner as an internal combustion engine that uses LPG.

The first commercial use of fuel cells was in NASA space programmes to generate power for probes, satellites, and space capsules. Since then, fuel cells have been used in many other applications. Historically, fuel cells have been used for primary and backup power to commercial, industrial, and residential buildings and in remote or inaccessible areas. They are increasingly frequently used to power buses, fork-lifts, airplanes, boats, motorcycles, and submarines.

Fuel cells come in many varieties; however, they all work in the same general manner. They are made up of three adjacent segments: the anode, the electrolyte, and the cathode. Two chemical reactions occur at the interfaces of the three different segments.

At the anode, a catalyst oxidises the fuel, usually hydrogen, turning the fuel into a positively charged ion (proton), and a negatively charged electron. A specifically designed membrane lets ions pass through it, but the electrons cannot. The protons pass through the membrane, while the freed electrons travel through an external circuit, thus creating an electric current. The ions travel through the electrolyte to the cathode. Once reaching the cathode, the ions are reunited with the electrons and the electrons and protons then recombine with oxygen on the other side of the membrane, producing water (H_2O) or water vapour as a byproduct. The net result of the two reactions is that the fuel (hydrogen) is consumed, oxygen from the air is consumed, water or water vapour is created, and an electric current is created, which can be used to power an electrical device (to recharge the vehicle battery or operate the electric motor), normally referred to as the 'load'.

To deliver the desired amount of energy, the fuel cells are combined in series and parallel circuits, called a fuel cell stack, where series yields higher voltage, and parallel allows a higher current to be supplied. The cell surface area is increased by design to allow stronger current from each cell.

FCEVs can have longer driving ranges than most battery electric vehicles, often in the range of 300–450 miles (500–700 km) per fill-up, and are refilled in a manner similar to LPG fuelled vehicles, in a time similar to that for refuelling a gasoline powered vehicle.

Hydrogen refuelling could be rolled out to the existing petrol stations, but would require substantial investment, and would have to be rolled out ahead of demand.

What then is the catch?

Currently, much of the hydrogen produced is produced using natural gas (not 'green' hydrogen extracted from water), which involves significant carbon emissions, therefore defeats the point.

Considerable efforts are underway to develop cleaner methods of hydrogen production (e.g. electrolysis using renewable energy, even to at some time in the future potentially using cyanobacteria to naturally produce hydrogen while at the same time removing CO_2 from the atmosphere).

The so-called 'green hydrogen' is hydrogen most commonly produced using renewable energy sources (such as wind, solar, or hydropower) through a process called electrolysis, which splits water into hydrogen and oxygen without emitting CO_2. Interest in green hydrogen is not just concerning FCEVs, but for power generation and industry as well.

The EU has ambitious plans to install large-scale electrolysis plants powered by renewable energy, with the aim of producing millions of tons of green hydrogen annually by 2030. Germany, Spain, and the Netherlands are leading in green hydrogen projects. Australia is leveraging its abundant solar and wind resources to become a green hydrogen export hub. Mega-projects such as the Asian Renewable Energy Hub are being developed to produce and export green hydrogen to markets such as Japan and South Korea, and Saudi Arabia and Morocco are investing heavily in green hydrogen production, using their vast solar and wind resources.

The following facilities are already under construction or in advanced design stage and committed to:

NorthH2 Project (Netherlands): Groningen, Netherlands.

- Electrolyser Capacity: 4 GW by 2030 (scalable).

A joint venture between Shell, Equinor, and other partners, NorthH2 will use offshore wind energy to power electrolysers in the Groningen region. The hydrogen will be used for industrial applications and distributed throughout Europe.

- Completion: Initial phase by 2027, with full capacity by 2030.

NEOM Green Hydrogen Project: NEOM, Saudi Arabia.

- Electrolyser Capacity: 2 GW.

This project is backed by a partnership between Air Products, ACWA Power, and NEOM. The plant will use 4 GW of solar and wind energy to produce green hydrogen. The project is expected to produce around 650 tons of hydrogen per day, which will be converted into ammonia for global export.

- Completion: Expected in 2026.

Hydrogen Utility (H2U) Eyre Peninsula Gateway Project: South Australia.

- Electrolyser Capacity: 75 MW (scalable to 1.5 GW).

H2U's Eyre Peninsula Gateway Project aims to create one of the largest green hydrogen production facilities in Australia. The project will initially produce green ammonia, with plans to scale up hydrogen production for domestic use and export. It will use wind and solar power to run the electrolysers.

- Completion: Initial phase by 2025.

HySynergy Project: Fredericia, Denmark.

- Electrolyser Capacity: 300 MW (scalable to 1 GW).

Developed by Everfuel and Ørsted, Phase 1 of the HySynergy project will focus on supplying hydrogen for heavy industry and transportation, with future expansion aimed at increasing capacity for broader green hydrogen use.

- Completion: Phase 1 by 2025.

Iberdrola Fertiberia Project: Puertollano, Spain.

- Electrolyser Capacity: 100 MW (scalable).

This project is a collaboration between Iberdrola and Fertiberia to use renewable energy to produce green hydrogen for use in fertiliser production. The initial phase features a 100 MW electrolyser powered by solar energy.

– Completion: Initial phase operational by 2025.

REFHYNE Project: Rhineland, Germany.

– Electrolyser Capacity: 10 MW (scalable to 100 MW).

REFHYNE is a European project led by Shell, ITM Power, and partners to produce green hydrogen for use in refining processes. The plant's scalable design plans to increase the initial capacity tenfold, marking it as a key element of Germany's hydrogen strategy.

– Completion: Initial 10 MW plant operational since 2021, with further expansion underway.

Green Hydrogen Hub: Eemshaven, Netherlands.

– Electrolyser Capacity: 1 GW (planned).

Developed by a consortium led by Shell, this green hydrogen hub will integrate offshore wind energy with large-scale electrolysers to produce green hydrogen. It is part of the Netherlands' broader goal to become a leader in hydrogen production and export.

– Completion: First phase by 2030.

The current process is well developed and reliable, but expensive. Whereas commercial hydrogen is available at circa US$ 1 per kg, Green hydrogen currently costs up to US$ 7 per kg to produce (mainly the cost of electricity),

This translates, for the FCEV, prior to taxes and duties, for example, using the Toyota Mirai (a current mid-sized sedan FCEV) as an example, which is claimed to have a hydrogen consumption rate of around 0.76 kg per 100 km under normal driving conditions and has a hydrogen fuel tank capacity of about 5.6 kg. To refuel with commercially available hydrogen, after distribution and retail mark up, would cost about US$ 10 before taxes and duty, whereas to refuel with 'green' hydrogen currently would cost US$ 50 before taxes and duty.

NOTE: fuel tax and duty policies vary greatly between countries, and cost per litre/kg are daily changing market prices, so these figures above and below, are only rough averages for comparison purposes.

With taxes and duties on fuel typically in the range of 100%–120% that gives an approximate comparison of US$ 100 against US$ 12–14.

For comparison this means that with a full (5.6 kg) tank of green hydrogen, the Toyota Mirai FCEV which has a range of over 700 km (450 miles) at an approximate cost of US$ 100, or US$ 14 per 100 km, or US$ 0.2 per mile. (These figures could be compared with an ICE car of similar size and achieving 35 mpg/8 litres per 100 km, consuming apx 56 litres of petrol costing US$ 110).

If the cost of producing green hydrogen can be reduced significantly (say halved), it becomes very cost competitive against gasoline/diesel ICE power, providing the commercial incentive to roll out into forecourts. And this could be done without disrupting the current tax/duty structure revenues in the way that the shift to BEVs will disrupt.

But governments have much greater ambition still. The global push to reduce the cost of green hydrogen is driven by ambitious targets set by governments and industries worldwide. In the short- to medium-term the US Department of Energy's Hydrogen Shot Initiative to reduce the cost of green hydrogen to $1 per kg by 2031 by advancing electrolyser technology, increasing the availability of low-cost renewable energy, and scaling up hydrogen production to achieve this cost target. The European Union strategy is to bring the cost of green hydrogen down to €1.50 per kg (around $1.60) by 2030 in regions with abundant renewable energy. The EU aims to install at least 40 GW of electrolyser capacity by 2030 and boost renewable energy production. Significant public and private investment is expected to accelerate the scaling of hydrogen projects.

For comparison this would mean that by the early 2030's a full (5.6 kg) tank of green hydrogen for the Mirai, with taxes and duties, would be around US$ 25, or US$ 4 per 100 km, or US$ 0.07 per mile. Far cheaper than gasoline, and therefore an attractive commercial roll-out.

In the longer term by 2040, it is expected that green hydrogen could reach cost parity with grey hydrogen in many parts of the world, even in regions with higher renewable energy costs. Industry forecasts suggest that in ideal conditions (regions with cheap renewable energy), green hydrogen costs could fall as low as $0.70 to $1 per kg by 2040–2050.

Confidence in achieving these targets is dependent on the mass production of electrolysers driving down capital costs. As more projects come online, the cost of electrolysers is expected to decrease by 50–70% by 2030, which combined with the decline in the cost of solar and wind energy (renewable electricity being the primary input for green hydrogen production) give confidence to proponents. Ongoing R&D into more efficient and durable electrolysers, new materials for catalysts, and alternative hydrogen production methods (e.g. advanced electrolysis technologies or biological hydrogen production) are expected to improve efficiency and further reduce costs further.

Major energy companies (such as Shell, BP, and Total) and industrial giants (such as Siemens and Mitsubishi) are investing in green hydrogen production as part of their decarbonisation strategies. Governments around the world are pledging billions of dollars to accelerate green hydrogen development.

In September 2024, BMW and Toyota announced a cooperation agreement to deepen their joint R&D efforts to develop hydrogen-based solutions for passenger cars.

Both companies have had ongoing hydrogen projects for more than a decade. BMW plans to launch its first production fuel cell electric vehicle (FCEV) in 2028 though it has been piloting hydrogen-powered cars since the late 1990s. Toyota has marketed the Mirai FCEV since 2014 (see below). Toyota and BMW have also been exploring liquid hydrogen as an alternative fuel, especially in motorsports (see below) and signed a previous agreement to collaborate in 2011. This new agreement extends that collaboration.

Toyota and BMW are now collaborating on the development of a third-generation fuel cell system, which will be installed in both companies' models to provide customers with a broader range of fuel cell electric vehicles (FCEV). As a first step, BMW plans to launch its first mass-produced FCEV in 2028.

Both companies will also work to create synergies through collaboration in development and procurement as well as drive down costs by amalgamating powertrain units to expand commercial and passenger vehicle demand. By making FCEV a more accessible option, the partnership will contribute to the creation of a hydrogen society. (source Toyota Newsroom).

Both companies believe that working with like-minded partners is essential for realising a hydrogen society. Given the need to create clusters of demand for hydrogen in the early stages of popularisation, Toyota and BMW will encourage sustainable hydrogen supply by creating demand and working closely with hydrogen-producing companies as well as distribution and refuelling facilities to ensure a stable hydrogen supply and to reduce costs.

Koji Sato, President of Toyota, said, 'We are pleased that the collaboration between BMW and Toyota has entered a new stage. In our long history of partnership, we have confirmed that BMW and Toyota share the same passion for cars and belief in "technology openness" and a "multi-pathway" approach to carbon neutrality. Based on these shared values, we will deepen our collaboration in efforts such as the joint development of next-generation fuel cell systems and the expansion of infrastructure, aiming for the realization of a hydrogen society. We will accelerate our efforts together with BMW and partners across various industries to realize a future where hydrogen energy supports society'.

Oliver Zipse, Chairman of the Board of Management, BMW AG, commented, 'This is a milestone in automotive history: the first-ever series production fuel cell vehicle to be offered by a global premium manufacturer. Powered by hydrogen and driven by the spirit of our cooperation, it will underscore how technological progress is shaping future mobility. And it will herald an era of significant demand for fuel cell electric vehicles'.

The companies say the joint venture will focus on shared powertrain technology 'utilized across individual models to offer attractive FCEV options'.

Alongside FCEV passenger cars, Toyota is applying its FCEV efforts to its Hino heavy truck unit because markets such as California have target dates to phase out diesel trucks altogether.

BMW's forthcoming X5-based FCEV successor to the iX5, currently offered to preferred customers for limited leasing as part of a development programme, due to launch in 2028, is the first in an extended lineup of BMW models to feature a new fuel cell stack which Toyota also is to employ in a follow-up to the Mirai sedan also planned for launch in 2028. (source Wards 100).

But the simple laws of physics determine that process of converting electricity to hydrogen and then using hydrogen to power vehicles or industry must always be less efficient than using electricity directly. This indicates that the true cost of electricity for BEVs should always be less than that for FCEVs. But the true costs of electricity are not the same as the market price. And the business model architecture may be more efficient for hydrogen.

If both FCEVs and BEVs have similar economies of mass production (at the moment BEVs have a significant advantage), the fact that an FCEV needs a battery of about 2 kWh whereas a typical BEV battery size ranges from 20 kWh to 100+ kWh, (depending on the vehicle) gives a manufacturing cost advantage to the FCEV, and the weight saved provides an operating efficiency advantage to the FCEV, offset by the complexity of their fuel stack equipment, and potentially higher maintenance costs.

FCEVs are not new, Fiat's fuel cell type vehicle Seicento Elettra H2 Fuel Cell, was introduced in 2001, followed by its Seicento Hydrogen, followed by the Fiat Panda Hydrogen (2005–2011).

The F-Cell is a hydrogen fuel cell electric vehicle developed by Daimler AG. Two different versions are known—the first version (2002) was based on the Mercedes-Benz A-Class, and the second (2010) model is based on the Mercedes-Benz B-Class. The first-generation F-Cell was introduced in 2002, and had a range of 100 miles (161 km), with a top speed of 82 mph (132 kmh). The 2010 B-Class F-CELL had a more powerful electric motor rated at 100 kW (134 hp), and a range of about 250 miles (402 km).

The Toyota Mirai is a mid-size hydrogen fuel cell electric vehicle (FCEV) manufactured by Toyota and was the first FCEV to be mass-produced and sold commercially. The Mirai was unveiled at the November 2014 Los Angeles Auto Show. As of November 2022, global sales totalled 21,475 units; the top-selling markets were the USA with 11,368 units, Japan with 7435 and the rest of the world with 2622. (source: Wikipedia).

Under the USA Environmental Protection Agency (EPA) cycle, the 2016 model year Mirai has a total range of 502 km (312 miles) on a full tank. The MPG-equivalent combined city/highway fuel economy rating was 66 mpg-US (3.6 L/100 km; 79 mpg-imp), making the Mirai the most fuel-efficient hydrogen fuel cell vehicle rated at the time by the EPA, and the one with the longest range. In August 2021, the second-generation Mirai set a world record of travelling 1360 km (845 miles) with a full tank of 5.65 kg (12.5 lb) of hydrogen. (Source: Wikipedia).

Hyundai Nexo: The Hyundai Nexo is an FCEV crossover SUV that offers a range of about 600 km (370 miles) on a full tank of hydrogen.

Honda Clarity Fuel Cell: The Honda Clarity Fuel Cell is part of Honda's Clarity lineup and offers a driving range of approximately 360 miles (580 km).

Summary

FCEV technology has been around for some time, is stable and efficient. The cars are very similar to internal combustion engine vehicles in their behaviour, range and refuelling, and generally have better economy than their ICE equivalent. Potentially more expensive to maintain than a BEV, but without the recharging delays and weight problems associated with BEVs. As Toyota Chairman Takeshi Uchiyamada said when announcing the intention to launch the Toyota Mirai FCEV 'The current capabilities of electric vehicles do not meet society's needs, whether it may be the distance the cars can run, or the costs, or how it takes a long time to charge'.

But the Achilles heel of the FCEV is the lack of availability of low-cost green hydrogen, leading to a lack of incentive to roll out the delivery infrastructure to service stations. As has been seen above, that should change by 2030–2035, and could be accelerated with incentives, but until then, and until there is infrastructure roll-out, FCEVs do not offer a workable solution for most.

Liquid Hydrogen ICE (HICEV)

A hydrogen internal combustion engine vehicle (HICEV) is simply a modified version of the traditional gasoline-powered internal combustion engine. HICEVS are different from hydrogen fuel cell vehicles (which utilise hydrogen electrochemically rather than through combustion).

In 1806, Francois Isaac de Rivaz designed the 'De Rivaz' engine, the first internal combustion engine, which ran on a hydrogen/oxygen mixture. Étienne Lenoir produced the 'Hippomobile' in 1863. Between 2005 and 2007, BMW tested a luxury car named the 'BMW Hydrogen 7', powered by a hydrogen ICE, which achieved 301 kmh (187 mph).

Back in 2001, 'Hydrogen Use in Internal Combustion Engines' by the US Department of Energy concluded 'Such as current gasoline-powered vehicles, the design of each hydrogen-powered vehicle will most likely vary from manufacturer to manufacturer and model to model. One mode may be simple in design and operation, for example, a lean burning fuel metering strategy using no emission control systems such as EGR, catalytic converter, evaporate fuel canister, etc. Another model may be very sophisticated in design and operation, for example, using an EGR fuel metering strategy with a catalytic converter, multiple spark plugs, etc.'.

Hydrogen internal combustion engine development has been receiving interest again recently, particularly for heavy-duty commercial vehicles. Part of the motivation for this is as a bridging technology to meet future climate CO_2 emission goals, and as technology more compatible with existing automotive knowledge and manufacturing.

In May 2023, Yamaha, Honda, Kawasaki, and Suzuki received approval from Japan's Ministry of Economy, Trade and Industry (METI) to form a technological research association called HySE (Hydrogen Small mobility and Engine technology) for developing hydrogen-powered engines for small mobility.

The hydrogen combustion engine has a peak at high load and can achieve similar efficiency levels as a hydrogen fuel cell. Hydrogen combustion engines are a match in terms of efficiency for fuel cells for heavy-duty applications, but efficiency decreases for small internal combustion engines.

In May 2023, Toyota Corolla Sport which is equipped with liquid hydrogen engine entered and completed the 'NNAPAC Fuji SUPER TEC 24 Hours Race'.

In the Netherlands, research organisation TNO (Netherlands Organization for Applied Scientific Research) has been working on the development of hydrogen internal combustion engines (ICEs) in collaboration with industrial partners.

TNO's research into hydrogen internal combustion engines focuses on leveraging existing ICE technology while transitioning towards more sustainable fuels. Hydrogen can be used as a clean-burning fuel, potentially offering a significant reduction in CO_2 emissions compared to traditional fossil fuels. TNO's work typically involves optimising engine design, improving fuel injection systems, and ensuring safe and efficient combustion of hydrogen. The goal is to offer an intermediate step towards full electrification by repurposing existing internal combustion technology to run on hydrogen, thus contributing to the energy transition. This initiative aligns with the Netherlands' and the European Union's broader goals for carbon reduction and the adoption of hydrogen as part of their energy strategy.

As can be seen in the example of the Toyota Corolla entered in, and completing, the 'NNAPAC Fuji SUPER TEC 24 Hours', Toyota has also been experimenting with liquid hydrogen as a fuel source for internal combustion engines, with the intent of reducing carbon emissions while maintaining a combustion engine design. Toyota consider that HICEV has the advantage that the technology retains much of the mechanical structure of conventional internal combustion engines, allowing Toyota to leverage decades of expertise in engine design and manufacturing and that liquid hydrogen has a higher energy density compared to gaseous hydrogen, which could potentially allow for greater range or more compact storage in vehicles.

The combustion of hydrogen with oxygen produces water vapour as the primary byproduct, rather than carbon dioxide (CO_2), making it a clean fuel option. But air is a mixture of gases, and the most abundant gas in air is nitrogen. Therefore, the combustion of hydrogen in air produces oxides of nitrogen, known as NOx. This problem is exacerbated by the very high temperatures generated by the combustion of hydrogen. This combustion process is much like other high temperature combustion fuels, such as kerosene, gasoline, diesel, or natural gas. Therefore, hydrogen combustion engines are not considered zero-emission vehicles. Toyota must therefore find ways to minimise NOx emissions to ensure compliance with environmental regulations.

There are other challenges as well. Liquid hydrogen must be stored at extremely low temperatures (around −253 °C or − 423 °F) to remain in liquid form, which requires advanced insulation and cryogenic technologies and the refuelling infrastructure for liquid hydrogen is virtually non-existent at present, and more expensive to install and maintain because of these insulation and cryogenic issues—making it difficult to deploy this technology at scale.

However, Toyota have just (October 2024) announced trials of *Hydrogen Cartridges* which they claim offer portable, affordable, and convenient energy that makes it possible to bring hydrogen to where people live, work, and play, without the use of pipes.

Prototype dimensions: 400 mm (16″) long x 180 mm (7″) diameter; Target weight: 5 kg (11 lbs).

The feasibility of this concept is supported by the lightweight nature of hydrogen. A litre of liquid hydrogen weighs just 0.0708 kg (0.156 pounds), meaning that a 140-litre tank, such as the one in the previously mentioned Toyota Mirai, would weigh around 10 kg (22 pounds) while providing a range of approximately 653 km (405 miles). This means drivers could easily transport enough hydrogen to drive many hundreds of miles by carrying a spare cartridge or two.

(Moreover, these portable tanks could also power household appliances, such as a hydrogen stove. BBQ, portable heater, etc. broadening hydrogen's potential uses beyond vehicles.)

Toyota claim that the cartridges can be swapped for easy replacement and quick recharging; that volume flexibility allows for a broad variety of daily use applications; that small-scale infrastructure can meet energy needs in remote and non-electrified areas and can be swiftly dispatched in disaster scenarios. Toyota recently showcased a swappable hydrogen fuel cell for electric vehicles at the Japan Mobility Bizweek.

Details of the financial viability of the solution have yet to emerge, and further study of safety aspects studied. But if the supplies of low-cost green hydrogen (discussed above) materialize, this cartridge solution could offer a viable alternative to BEVs in the future.

Butanol

Butanol can be manufactured through two primary processes: biological fermentation and chemical synthesis. Butanol by chemical synthesis is discounted because it uses petrochemicals as feedstock.

The biological fermentation process 'acetone-butanol-ethanol (ABE)' uses microorganisms, typically 'Clostridium acetobutylicum', to convert carbohydrates (sugars, starches, or lignocellulosic biomass) such as corn, sugarcane, or agricultural residues. Microbes break down the sugars in the feedstock under anaerobic conditions, producing butanol, acetone, and ethanol. The typical ratio is 6 parts butanol, 3 parts acetone, and 1 part ethanol. The fermentation broth is distilled to separate and purify the butanol from other products. Solvent extraction or gas stripping can be used to recover butanol efficiently during the fermentation.

This process is renewable and relies on bio-based materials, but requires the diversion of land from food production, and the farming process also generates emissions.

New methods, such as genetically engineered microbes and more efficient catalytic processes, are being researched to make butanol production more sustainable and cost-effective.

When used as a fuel, butanol has advantages over ethanol, such as higher energy content and compatibility with gasoline engines without significant modifications.

But because of the amount of agricultural land that would be required, is not further considered by this work as a practical alternative to BEVs (although it may be a contributor to reducing ICEV emissions).

Chapter 6
Other Issues

Abstract Chapter 6 addresses open issues, returning to look more deeply at the problem of weight and impact on roads and infrastructure and developments in lightweight BEV batteries. The chapter then addresses other issues such as the problem of accident and insurance, the 'rare earths' issues, the taxation issues, the payment for charging issues and asks the question 'Just how green is a current BEV?

Keywords Solid-state batteries · Structural batteries · Lithium-ion battery fires · Rare earth element · Environment issues · Distance-based road user charging · ZEV Mandate

The Problem of Weight

BEVs face several significant challenges related to their weight.

The first problem is that a BEV battery pack is heavy in comparison to the rest of the weight of the vehicle, so it requires more powerful motors to propel the heavy vehicle. More powerful motors require more power which requires bigger batteries which add further to the weight, requiring even more powerful motors, and so on.

By the time the manufacturer has optimised, the battery pack is often accounting for 25% to 35% of the BEV's total weight. For instance, a typical BEV battery pack can weigh anywhere from 300 to 600 kg (660 to 1320 lbs). The situation for BEVs designed for longer ranges is even more acute, requiring even larger (heavier) batteries, which adds even more weight. For example, a long-range electric vehicle such as a Tesla Model S or Model X may have a battery weighing around 540 kg (1200 lbs) or more.

BEVs are therefore significantly heavier than their ICE counterparts due to the weight of the battery pack. For example, a Tesla Model 3 weighs around 1600–1900 kg, whereas a similar-sized ICE vehicle may weigh 1300–1400 kg, some 25%–30% extra weight.

The heavier the vehicle is, the more energy it takes to move it. This can create a diminishing returns problem where adding more battery capacity to extend range

B. Williams, *Electric Shocks: Are Drivers Rejecting the Electric Car?*,
https://doi.org/10.1007/978-3-031-98820-2_6

increases the vehicle's weight, which in turn increases the energy needed to propel the vehicle, reducing the efficiency gains.

In general, heavier vehicles accelerate less rapidly and require longer stopping distances. While BEVs often compensate for this with more powerful electric motors, the increased weight still affects overall performance, particularly in handling and braking. The vehicle's suspension system and tyres are under greater stress, leading to increased wear and potentially reducing their lifespan. BEVs are designed with reinforced suspension systems and use heavier tyres to manage the extra weight.

A heavier BEV requires more energy to accelerate and maintain speed, particularly at higher speeds and during uphill driving. This results in higher energy consumption which reduces the effective driving range.

Heavier vehicles can be safer for their occupants in a crash because the added mass can absorb more impact. But the increased weight can exacerbate the risk of battery damage in severe accidents, which can cause fires or other complications.

Impact on Roads and Infrastructure

Heavier vehicles cause more wear and tear on roads (see Fig. 6.1), bridges (see Fig. 6.2), and parking structures.

While BEVs are generally lighter than large commercial vehicles, their added weight compared to ICEVs accelerates the deterioration of public infrastructure over time.

Fig. 6.1 Potholed road: source—CSi UK Ltd. Library

Fig. 6.2 Bridge weight restriction: source—CSi UK Ltd. Library

The additional weight of BEVs (and SUVs) can also pose challenges for parking structures (see Fig. 6.3).

When driving in urban environments, where stop-and-go driving is common, the weight problem can make BEVs less energy-efficient because it takes more energy to start and stop a heavier vehicle repeatedly.

The weight problem of BEVs is caused entirely by the heavy battery packs required to power them. Auto manufacturers are acutely aware of this problem and several innovations show promise.

Developments in Lightweight Battery Electric Vehicle (BEV) Batteries

Solid-State Batteries: As discussed briefly in Chap. 4 (section "Recent and Ongoing Battery Developments"), solid-state batteries replace the liquid or gel electrolyte in conventional lithium-ion batteries with a solid electrolyte, which can be made of ceramics, polymers, or other solid materials. These batteries offer higher energy density, meaning more energy can be stored in a smaller and lighter package. Solid-state batteries are also safer, with a significantly reduced risk of fire or thermal runaway. They could also enable BEVs to have longer ranges without increasing weight.

Companies such as Toyota, Idemitsu Kosan, QuantumScape, and Solid Power are leading the way in solid-state battery development. Toyota, for instance, plans to

Fig. 6.3 Partially closed car park because of structural issues: source—CSi UK Ltd. Library

commercialise solid-state batteries in the near future, targeting lighter, more efficient BEVs.

Idemitsu Kosan Co. Ltd. (Idemitsu) and Toyota Motor Corporation (Toyota) announced in June 2023 that they have entered into an agreement to work together in developing mass production technology of solid electrolytes, improving productivity and establishment a supply chain, to achieve the mass production of all-solid-state batteries for BEVs. Through this collaboration, the two companies, which lead the world in the fields including material development relating to all-solid-state batteries, seek to ensure the successful commercialisation of all-solid-state batteries in 2027–28, followed by full-scale mass production.

In September 2024, Yu Jingmin, executive deputy general manager of SAIC Passenger Vehicle, stated in a media interview that SAICs self-developed all-solid-state battery will be first installed on an MG car. It is believed the vehicle will be launched in the second quarter of 2025. ‘The main problem with making solid-state batteries commercially viable is being able to produce them cheaply enough through mass production.’.... ‘Initially, such batteries will be exclusive to more high-end models and competing against NCM chemistry batteries.’

Xu Yanhua, Secretary-General of the China Automotive Power Battery Industry Innovation Alliance, stated ‘Before 2030 the battery market in China will still be dominated by liquid batteries.’

Lithium-Sulphur (Li-S) Batteries: As discussed briefly in Chap. 4 (section "Recent and Ongoing Battery Developments"), lithium-sulphur batteries use sulphur in the cathode instead of the traditional nickel-cobalt-manganese materials used in lithium-ion batteries. Sulphur is lighter, more abundant, and cheaper than materials used in current lithium-ion batteries. Lithium-sulphur batteries could offer up to five times the energy density of lithium-ion batteries, drastically reducing weight while maintaining or improving range. The primary challenge has been the short lifespan of Li-S batteries due to sulphur's tendency to degrade rapidly. However, advancements are being made to improve the cycle life. Companies such as Oxis Energy and research institutions are working to bring this technology closer to commercialisation.

Lithium-Air (Li-Air) Batteries: Lithium-air batteries use oxygen from the air as the cathode, which greatly reduces the weight of the battery compared to traditional lithium-ion designs. These batteries have the potential to achieve energy densities, offering the possibility of extremely lightweight and long-range BEVs.

Significant technical challenges remain, including the instability of the chemical reactions and difficulty with recharging, so do not expect commercialisation in the near future.

Silicon Anodes: As discussed briefly in Chap. 4 (section "Recent and Ongoing Battery Developments"), silicon is being explored as a replacement for graphite in battery anodes. Silicon can store significantly more lithium ions than graphite, allowing for higher energy density. Using silicon in the anode could increase the energy density of batteries by as much as 20–40%, making the batteries lighter while maintaining the same capacity or allowing for greater capacity without a weight penalty.

But silicon expands and contracts significantly during charging and discharging, which can cause the material to break down over time. Companies such as Sila Nanotechnologies and Amprius are working on advanced silicon anode technologies that address these durability issues.

Structural Batteries: Structural batteries integrate the battery into the actual structure of the vehicle, such as the body panels or chassis, rather than housing them in a separate pack. This approach reduces weight by combining two functions—energy storage and structural support—into a single component. Structural batteries could significantly reduce the overall weight of BEVs by eliminating the need for a traditional battery pack casing and reducing the weight of the body itself.

Researchers at institutions such as Chalmers University in Sweden and companies such as Tesla have been exploring structural batteries. Tesla's recent innovations in battery pack design, such as their '4680' cells and structural battery packs, are steps towards achieving this goal.

Graphene Batteries: Graphene-based batteries use graphene, a material made from a single layer of carbon atoms, to improve conductivity and energy storage. Graphene batteries could offer higher energy density, faster charging times, and longer life cycles while being lighter than traditional lithium-ion batteries.

Companies such as Skeleton Technologies and research groups are developing graphene-enhanced batteries. While still largely in the research phase, graphene batteries hold promise for lightweight, high-performance BEVs.

Conclusion

These developments in battery technology aim to address the weight problem of BEVs by increasing energy density, improving efficiency, and using new materials and designs to reduce the overall weight of the vehicle. As these technologies advance and are commercialised, they could lead to lighter BEVs with longer ranges, better performance, and lower costs, all while maintaining safety and sustainability. The problem is that, at the moment, these are largely 'research' projects that 'could' solve the weight problems of BEVs, and as anyone who has been on a diet to lose weight will tell you—it is much easier to talk about weight loss than it is to actually achieve it!

For the moment, BEVs, and the infrastructure they encounter, are stuck with their weight problem and its consequences.

The Problem of Accident and Insurance

When a gasolene powered vehicle is involved in a crash, the emergency responder's first concern, normally even above that of rescuing the occupants, is to prevent explosion, or prevent or extinguish any fire associated with or threatening its highly flammable fuel supply. But that done, assuming the fuel tank and supply lines are undamaged or replaced, there is no long-term effect on the lifetime of the vehicle, whose future depends on the cost of repairing the mechanical and any electronic damage to the vehicle. If necessary, replacing the fuel tank and supply line pipes will not normally be a material factor in those decisions.

When a BEV is involved in a crash, the emergency responders' first priority is to disable the battery and get the occupants out and away from the vehicle before the battery explodes or a lithium induced fire erupts. We learned in Chap. 4 about the risks of lithium and cobalt fires (see section 'The Challenges of Working with the Raw Materials of BEV Batteries' in Chap. 4 above). Lithium-ion battery fires are more difficult to extinguish than normal car fires and require specialised firefighting techniques. Insurers may view this as a potential risk factor, especially in the context of post-accident fires, which can complicate claims and raise premiums.

But unlike the ICE vehicles fuel tank, any damage to the BEVs battery pack is crucial to the future of the vehicle. The BEVs battery pack represents 35%–45% of the cost of the vehicle, so, taken together with mechanical and bodywork repairs required, the need for the replacement of the BEV battery pack is crucial in determining whether the vehicle is written off.

And further, given the risks of explosion or fire associated with exposure to water, air, and heat, who is able to say with confidence, following a crash, that the sealed battery pack has not incurred some damage that may cause its lithium or cobalt to be exposed to air or water at some time over the rest of its working life? It is not just a question of a quick visual check and a test to see that the battery is still working. Insurers are about risk minimisation (to them) and have shown a tendency to just write the vehicle off.

The battery pack is the most expensive component in a BEV. In the event of a collision, even minor damage to the battery can lead to costly repairs or, for the reasons given above, lead to a full battery replacement, or cause a write-off. This can significantly impact insurance premiums.

Repairing BEVs often require specialised mechanics and facilities for repairs, particularly for their complex electrical systems and high-voltage BEV batteries. This can result in higher labour costs and there are fewer repair shops capable of servicing BEVs (who can command higher prices), contributing to more expensive claims, leading to more expensive insurance premiums.

Unless subsidised, BEVs have significantly higher initial purchase prices than comparable ICE vehicles, generally around a third more expensive (largely due to the cost of the battery). As a result, the replacement cost in the event of total insurance loss is higher, which again is a cause of higher insurance premiums.

And some auto manufacturers are still subsidising the price of BEVs, particularly smaller models, as part of their marketing strategy to gain wider acceptance. Meaning that significant repairs can often exceed the vehicle cost, again leading insurers to write off rather than repair.

As a further disadvantage, BEVs generally depreciate faster than traditional vehicles due to concerns about battery life and technological obsolescence. Insurers may factor in this depreciation when determining payouts for claims, affecting coverage terms.

Many BEV owners install charging stations at home. Insurers may offer additional coverage for damage to or theft of home charging equipment, but this could also increase the overall cost of the policy.

BEV drivers may face unique risks related to public charging stations, such as theft of charging cables, damage while charging, or vandalism. Insurance policies may need to address these risks.

Emergency Assistance: BEV drivers may require specialised roadside assistance, particularly in cases where they run out of charge. Towing a BEV may require specialised equipment due to the weight of the vehicle and the need to avoid damaging the battery. Insurance companies may need to offer tailored roadside assistance plans for BEV owners.

Environmental Liability: Although BEVs are viewed as being more environmentally friendly, in the event of a crash they present unique environmental risks, especially related to battery disposal. In the event of a severe crash, insurers might be responsible for addressing environmental cleanup related to damaged or leaking batteries, increasing potential liability, and consequently raise premiums.

Insurance costs for BEVs can vary depending on the region. Areas with more extensive charging infrastructure and a higher number of qualified repair shops may have more competitive insurance options. In contrast, rural or less-developed areas may face higher premiums due to the lack of specialised services.

Conclusion

BEVs introduce unique challenges for insurers. Higher purchase costs, higher repair costs, battery-related risks, and uncertainties around residual value and data availability all contribute to more expensive premiums. However, as the market matures and more data becomes available, insurance products are likely to become more tailored and competitive for BEV owners. At the moment, on like-for-like market value, BEV insurance premiums are likely to be up to 20% more expensive than their ICEV equivalent. But as the market cost of the BEV is about a third more than its ICE equivalent, the true difference that hits the driver's pocket is likely to be greater.

The 'Rare Earths' Issues

BEVs typically use permanent magnet synchronous motors (PMSMs) which depend on rare earth elements to create the powerful permanent magnets that are central to their design. Permanent magnets enable motors to have a higher power density and greater efficiency, making them ideal for BEVs where size, weight, and energy efficiency are critical. Rare earth magnets allow for smaller, lighter motors that still produce high torque, which is crucial for performance and range in BEVs.

A so-called 'rare earth element' (REE) refers to any of a group of 17 chemical elements in the periodic table. These elements are categorised into a grouping named the 'lanthanides', along with scandium and yttrium (which exhibit similar properties and are often found in the same mineral deposits). The full list of rare earth elements is as follows:

1. Lanthanides (15 elements):
   - Lanthanum (La)
   - Cerium (Ce)
   - Praseodymium (Pr)
   - Neodymium (Nd)
   - Promethium (Pm)
   - Samarium (Sm)
   - Europium (Eu)
   - Gadolinium (Gd)
   - Terbium (Tb)

  - Dysprosium (Dy)
  - Holmium (Ho)
  - Erbium (Er)
  - Thulium (Tm)
  - Ytterbium (Yb)
  - Lutetium (Lu)

2. Scandium (Sc)
3. Yttrium (Y)

Scandium and yttrium are considered to be rare earth elements (REEs) because they tend to occur in the same ore deposits as the lanthanides, and they exhibit similar chemical properties (but they have different electrical and magnetic properties).

All rare earth elements have similar properties, especially in their electronic structure, which makes them useful in specific high-tech applications, but they behave/react slightly differently. Commonly, they are typically silvery-white, soft metals that tarnish when exposed to air. These metals tarnish slowly in air at room temperature and react slowly with cold water to form hydroxides, liberating hydrogen. They react with steam to form oxides and ignite spontaneously at a temperature of 400 °C (752 °F).

These elements and their compounds have no biological function other than in several specialised enzymes, such as in lanthanide-dependent methanol dehydrogenases in bacteria. REEs are often found in minerals with thorium, and less commonly uranium.

Some REEs, such as neodymium and dysprosium, are essential for manufacturing strong permanent magnets, such as the motors that power a BEV, while others, such as europium and terbium, are valuable for their luminescent properties.

Despite their name, rare earth elements are not actually rare in terms of abundance in the Earth's crust. Many REEs are relatively plentiful but they are rarely found in concentrated, economically extractable amounts. The term 'rare' originated in the eighteenth and nineteenth centuries when these elements were first discovered in rare minerals, leading to the use of the term 'rare'. But although rare earth elements are technically relatively plentiful across the entire Earth's crust (cerium being the 25th-most-abundant element at 68 parts per million and is more abundant than copper), in practice they tend to be spread very thinly across trace impurities. This means that in order to obtain rare earths at a usable purity requires that enormous amounts of raw ore have to be processed (at great expense), so the use of the term 'rare' earths is apposite. Because REEs tend to be spread very thinly, economically exploitable ore deposits are sparse.

The processing of REEs involves several stages, starting with mining and concentration of the ore, followed by separation and purification of individual REEs, and finally converting them into usable forms such as oxides or metals.

REEs are typically extracted from ores such as bastnaesite, monazite, and xenotime. These ores contain a mixture of REEs along with other minerals.

- **Mining:** The mining process for REEs involves either open-pit mining or underground mining, depending on the location and depth of the ore. (Open-pit mining is more common due to the nature of REE deposits). Following its extraction, the ore is crushed into small pieces to facilitate further processing. The crushed ore is subsequently ground into fine particles in order to liberate the rare earth minerals from the surrounding rock. This step increases the surface area for subsequent chemical processes. The next step is 'concentration', which is achieved by one of several methods:
- **Flotation:** Flotation is a common method used to concentrate the REEs by separating the REE-containing minerals from the waste material (gangue). In this process, the ground ore is mixed with water and chemicals that cause the REE minerals to become hydrophobic (repel water). Air bubbles are introduced, and the REE minerals attach to the bubbles, rising to the surface to form a froth, which is then skimmed off.
- **Magnetic Separation:** Magnetic Separation exploits the magnetic properties of certain REE minerals. The ore is passed through a magnetic field, which attracts the magnetic minerals, separating them from non-magnetic materials.
- **Electrostatic Separation:** Electrostatic Separation uses electrostatic forces to separate REE minerals based on their electrical conductivity. (It is particularly effective for separating non-conductive REE minerals from conductive impurities).
- **Gravity Separation:** Gravity Separation is sometimes used to concentrate on heavy rare earth elements (HREEs) from lighter minerals. The ground ore is passed through devices such as jigs, spirals, or shaking tables that exploit differences in density to separate REE minerals from lighter waste material.

Once separated, the next stage of the process is *Refining*. Once again there are multiple methods to achieve this step:

- **Acid Leaching:** The concentrated REE minerals are treated with strong acids (such as sulphuric or hydrochloric acid) in order to dissolve the rare earths into a solution in order to separate the REEs from the remaining non-REE minerals and other impurities.
- **Alkali Leaching:** The concentrated REE minerals are treated with alkalis to dissolve the rare earths into a solution in order to separate the REEs from the remaining non-REE minerals and other impurities.
- And is followed by:
- **Residual Processing:** Where the waste materials, often referred to as tailings, are usually neutralised and disposed of. (This can result in environmental concerns, particularly if radioactive elements are present.)

And then:

- **Solvent Extraction:** Solvent extraction is the most crucial and complex step in rare earth processing. It is used to separate individual rare earth elements from one another due to their similar chemical properties.

In one method to achieve this extraction, the leachate containing dissolved REEs is mixed with organic solvents in a series of stages. In each stage, different REEs preferentially transfer between the aqueous and organic phases, gradually separating the elements based on their slight differences in chemical behaviour. This process is repeated many times through multiple extraction stages, known as mixer-settlers, to achieve the desired purity of individual REEs.

Ion exchange is an alternative or complementary method to solvent extraction. It involves passing the REE-containing solution through a resin bed that selectively binds specific rare earth ions. The REEs are then eluted from the resin with a suitable reagent, separating them from other ions. This method is particularly useful for separating heavy rare earth elements (HREEs) from light rare earth elements (LREEs).

And finally, there are purification and final product formation processes:

- **Precipitation:** After separation, the individual rare earth elements are precipitated from the solution as oxalates, carbonates, or hydroxides by adding appropriate chemicals (e.g. oxalic acid, ammonium bicarbonate). The precipitated REE compounds are then filtered out, washed, and dried to remove remaining impurities.
- **Calcination:** The precipitated REE compounds are heated in a process called calcination, converting them into oxides. For example, REE carbonates or oxalates are heated to decompose them into their corresponding oxides.

The result is a purified rare earth oxide, which is the final product in most cases. These oxides are often sold as powders, which can be further refined or reduced to metals if needed.

In some cases, REE oxides are reduced to their metallic form using intensive high energy use high-temperature reduction techniques, such as electrolysis or achieve reduction with calcium or lanthanum metal. The metals are then used to produce alloys or other compounds used in various high-tech applications, including motor magnets and batteries.

Environmental and Waste Management Issues: The processing of REEs generates significant amounts of waste, including tailings that may contain radioactive elements, heavy metals, and other toxic byproducts. Proper management of these tailings is critical to avoid environmental contamination.

Acidic or alkaline waste solutions have to be neutralised before disposal to prevent damage to the environment. However, this process itself can generate large volumes of waste material that require careful handling.

Water used in the processing stages has to be treated to remove contaminants before being released back into the environment. This may involve sedimentation, filtration, and chemical treatment to remove dissolved metals and other pollutants.

Summary

Processing rare earth elements involves several stages, starting with mining and concentration of the ore, followed by separation and purification of individual REEs, and finally converting them into usable forms such as oxides or metals. The process is complex, energy-intensive, and often environmentally challenging, especially due to the management of radioactive waste and other byproducts. These challenges have led to significant environmental concerns and prompted efforts to develop more sustainable and less polluting processing technologies.

The process is complex, energy-intensive, and often environmentally challenging, especially due to the management of radioactive waste and other byproducts. These challenges have led to significant environmental concerns and prompted efforts to develop more sustainable and less polluting processing technologies.

The point of working through these processing stages is to show that although in weight or volume terms the amount of REEs used in the manufacture of a BEV (or its battery pack) is miniscule, it has consumed great quantities of energy in its manufacture. Where that energy is not green (and China, where 80% of REEs are processed has a target of 2060 before it is net-zero) it adds significantly to the carbon footprint of a BEV.

Uses of Rare Earth Elements

Rare earth elements play a critical role in many modern technologies, including BEVs, due to their unique magnetic, chemical, and optical properties:

- Magnets: Neodymium, praseodymium, and dysprosium are used in the production of powerful permanent magnets found in BEV motors, wind turbines, etc.
- Catalysts: Cerium and lanthanum are used in catalysts for refining petroleum and in catalytic converters in cars.
- Phosphors and Displays: Europium and terbium are used in phosphors for lighting and display screens, such as LEDs, LCDs, and smartphone displays.
- Lasers and Optics: Certain REEs are used in high-performance lasers and other optical technologies.

Supply and Geopolitical Concerns China and Russia control a significant portion of the world's rare earth production and refining capacity, which has raised geopolitical concerns about supply chain dependencies. Other countries, such as the USA, Australia, and Japan, are investing in rare earth mining and processing to reduce reliance on China and Russia.

The extraction and processing of rare earth elements can cause significant environmental damage, including soil degradation, radioactive waste, and water contamination, especially when not properly managed.

Rare Earth Elements and BEVs

BEV batteries rely on several critical materials, including some that are classified as rare earth elements (REEs) and critical minerals, such as lithium, cobalt, nickel, and others. The extraction, processing, and supply of these materials pose significant environmental, ethical, and geopolitical challenges.

The issues concerning lithium, cobalt, nickel, and other critical minerals have already been discussed in the chapters above; this section focuses on BEVs and REEs.

BEV manufacturing involves the use of rare earth elements (REEs) in key components, especially in the motors and electronics that enable the vehicle to function. Rare earth elements such as neodymium, dysprosium, and praseodymium are used in the permanent magnets of electric motors found in BEVs, rather than in the batteries themselves. But some battery architectures involve the use of REEs.

The use of rare earth elements introduces several challenges, including supply chain dependency, environmental damage, and ethical concerns. The most relevant rare earth elements in BEVs include:

- Neodymium (Nd): Primarily used in the production of high-performance permanent magnets for electric motors.
- Praseodymium (Pr): Often used in conjunction with neodymium to enhance magnetic properties.
- Dysprosium (Dy): Used to improve the heat resistance of permanent magnets, making them more stable in high-temperature environments (important for electric motors in BEVs).
- Terbium (Tb): Also used to improve the performance of magnets.
- Lanthanum (La) and Cerium (Ce): Used in certain battery types and in catalysts, though these are more common in hybrid vehicles rather than BEVs.

BEVs typically use permanent magnet synchronous motors (PMSMs), which are highly efficient and lightweight. These motors depend on rare earth elements such as neodymium, praseodymium, and dysprosium to create the powerful permanent magnets that are central to their design. Permanent magnets enable motors to have a higher power density and greater efficiency, making them ideal for BEVs where size, weight, and energy efficiency are critical. Rare earth magnets allow for smaller, lighter motors that still produce high torque, which is crucial for performance and range in BEVs.

Permanent magnets containing REEs are also used in regenerative braking systems and various electronics within BEVs. Lanthanum and cerium are used in certain display and lighting technologies; phosphors that rely on REEs such as europium and terbium are used in vehicle displays, LEDs, and lighting systems.

The global rare earth element supply chain is highly concentrated, with China dominating both the mining and processing of these elements. According to estimates, China controls over 80% of the world's rare earth processing capacity. This concentration creates vulnerability to supply disruptions, trade restrictions, and political leverage. For example, China has used its control of rare earth supplies as a bargaining chip in international disputes, notably with the USA and Japan.

The monopolisation of rare earth processing leads to potential price volatility, which can impact BEV production costs. A sudden price spike in rare earths could drive up the cost of BEV motors and other components, making vehicles more expensive for consumers.

Even though rare earth elements can be found in various parts of the world, their extraction is only one part of the problem. Processing these elements to a form usable in high-tech applications, such as BEV motors, is a complex and capital-intensive process. The lack of processing facilities outside China adds to the supply chain risks.

Governments and companies are working to diversify the rare earth supply chain by developing new mining and processing capabilities in other regions: the USA, Australia, and Canada are increasing investments in rare earth mining and processing to reduce dependence on China. New mines are being developed in these countries, and processing capacity is being expanded.

Mining rare earth elements can generate radioactive waste, soil erosion, and water contamination, especially when not properly managed.

The mining, refining, and processing of critical materials for BEVs, including REEs, are highly energy-intensive processes, often relying on fossil fuels. This can offset the claim of BEVs to have a zero-carbon footprint, or at least a smaller carbon footprint than its ICEV equivalent, especially if the electricity grid in the production region is not renewable.

The use of rare earth elements in BEV manufacturing presents both opportunities and challenges. While they enable the production of highly efficient electric motors and other essential components, the sourcing and processing of these materials come with significant carbon emission, environmental, ethical, and geopolitical issues. Addressing these concerns will require a combination of technological innovation, improved recycling systems, and a diversification of the global supply chain.

The Taxation Issues

Fuel duties—taxes—on fuel for vehicles (gasolene/petrol and diesel), whether as a quantitative duty per litre/gallon, or a tax on the value of the purchase transaction (VAT or sales tax), or most commonly a cumulative combination of the two, is a significant generator of income for governments around the world.

For example in the UK approximately 2%–3% of government revenues are from fuel taxes (53p per litre) plus 20% VAT on the combined price of the fuel plus duty, which compounds to about 3.5%–4% of its total revenue. In France approximately 3%–5% of government revenues from fuel taxes, compounding to 5.5%–6% including VAT; Germany 4.5%–5.5%. (The USA varies from state to state, and is less onerous, but still onerous—fuel taxes in California for example are some US$ 0.78 per gallon.).

If citizens around the world do as encouraged—or instructed by Regulation—convert to driving BEVs, how will Governments make good the shortfall of 3.5%–5.5% of their income?

If income tax were to be raised to compensate, that would imply raising the basic rate of income tax by 7p–11p in the pound/euro/dollar—a rise of 35% or more.

Pay-per-Mile/Kilometre

An alternative proposed is a pay-per-mile/kilometre-driven tax. But many countries already charge road tolls for the use of motorways/autoroutes. This would imply double taxation when using tolled highways.

Pay-per-mile road taxation, also known as distance-based or mileage-based road user charging, is a system where drivers are taxed based on the number of miles/kilometres they drive rather than through traditional fuel taxes or flat road taxes. While this system has some advantages, it also comes with several potential disadvantages:

Rural vs. Urban Drivers: Rural drivers, who typically need to drive longer distances due to fewer public transport options and greater distances between destinations, may be disproportionately affected by pay-per-mile taxes. This could make the system seem unfair to those who live in rural areas.

Households with lower incomes might be more negatively impacted if they need to drive long distances for work or other necessities, exacerbating financial burdens.

Consider, for example, what is fair about a tax which will be paid more by those on lower wages who must go out to earn a living, such as low paid domiciliary care workers and delivery drivers, than by wealthy professionals or public sector workers who can opt to work from home?

For the metropolitan elite, pay per mile makes sense. But it is the urban mileage that creates greater pollution rather than the longer in distance across country commute. Penalising individuals for their unavoidable travel distances is claimed to be unfair and hit low-income families the hardest. But you can argue that taxing per litre is no different. Except that, unless the tax rate is linked to the power or cost of the vehicle, charging per mile is regressive on small family runabouts. If linked to cost of the vehicle, is regressive on BEVs which are more expensive, discouraging the switchover to BEVs.

Pay-per-mile systems require tracking the number of miles driven, which can involve GNSS tracking or other monitoring technologies. This raises significant privacy concerns, as it could allow the government or third parties to monitor individuals' driving patterns, locations, and movements. There is also a risk of data breaches where sensitive information about drivers' locations and travel habits could be exposed or misused.

Setting up a pay-per-mile system would be expensive. It requires the installation of monitoring devices in vehicles, development of the necessary infrastructure to collect and process the data, collect the fees and ongoing maintenance and support systems. Equipping the existing car park is a particularly difficult problem to overcome.

The administrative costs associated with managing a pay-per-mile system, including billing, enforcement, and customer service, is likely to be higher than traditional tax systems.

Unlike fuel taxes, which naturally penalise less fuel-efficient vehicles (since they consume more fuel per mile), a pay-per-mile system does not directly encourage drivers to choose more fuel-efficient vehicles. This could undermine efforts to reduce carbon emissions and improve overall vehicle efficiency.

The technology used for tracking mileage must be highly reliable. Technical failures or inaccuracies in tracking could lead to incorrect billing, disputes, and loss of public trust in the system. However, technical accuracy is not a problem as several technical solutions exist and pay-per-mile insurance has been operating successfully for a number of years. Public acceptance will be a much bigger issue.

Establishing who pays in the case of vehicles with multiple drivers is an issue, but governments will probably charge the registered keeper of the vehicle, and let the registered keeper sort those issues out.

The technical means of payment will be another issue as the sums of money may be very significant, especially for the lowest income quartile. Governments will prefer 'direct debit' type arrangements, civil rights organisations and regulators may oppose.

Ensuring all drivers comply with a pay-per-mile system can be challenging. There could be issues with drivers attempting to tamper with or disable tracking devices, leading to enforcement difficulties.

But the big issue will be public acceptance. Many drivers are accustomed to paying taxes through fuel purchases, and there may be significant public resistance to a new, potentially more intrusive system of taxation. If the public perceives the pay-per-mile system as unfair or overly complex, there may be strong opposition, making it difficult to implement and sustain.

Transitioning from fuel taxes to a pay-per-mile system will be complex and may involve a period of double taxation or confusion over how taxes are applied.

While pay-per-mile road taxation offers a potential solution to there being no use of gasolene/diesel to tax, it comes with significant disadvantages, including privacy concerns, high implementation costs, equity issues, and potential public resistance. These factors must be carefully considered and addressed to create a system that is effective, fair, and acceptable to the public.

Change Annual Vehicle Tax to a Monthly (or Weekly) Road Use Tax

Assuming that the taxation level would be adjusted according to the characteristics or value of the vehicle, this would be technically simpler and less obtrusive to implement. It would also be far, far cheaper to implement and maintain.

But charging a flat rate per vehicle, even if adjusted according to the characteristics/price of the vehicle is regressive on low-mileage vehicles. If linked to cost of the vehicle, is regressive on BEVs which are more expensive, discouraging the switchover to BEVs. Furthermore, it does nothing to encourage people to drive less miles.

The scale of such charging would also be very significant, probably in the range of £150–200/€175–€240/US$190–$260 per calendar month.

Taxing Use of Electricity for Vehicles

It is technically possible. In a scenario where all cars are BEVs, public charging points could easily be taxed and charging prices adjusted. Home charging points could be fitted with a smart meter and the consumption taxed (but an expensive solution to retrofit); and how the system would deal with trickle charging via a domestic mains plug or solar panels, would have to be worked out.

Another solution would be to meter electricity consumption in the vehicle. That solution would involve a little new technology in the vehicle (not difficult but would mean a design change for every vehicle on the market); and a new and reliable system and equipment to report this to the taxation authorities (novel and expensive unless connected vehicle V2X technology is enforced).

Setting up a pay-per-vehicle-usage-of-electricity (PPVUE) system would be expensive. It requires the installation of monitoring devices in vehicles, development of the necessary infrastructure to collect and process the data, collect the fees and ongoing maintenance and support systems. Equipping the existing car park is a particularly difficult problem to overcome.

The administrative costs associated with managing a PPVUE system, including billing, enforcement, and customer service, is likely to be higher than traditional tax systems.

Increase VAT/Sales Tax

The simplest and bluntest way to replace this loss of income would be to increase VAT. This could be justified on the grounds that little of the existing fuel taxes raised is spent on the road infrastructure. However, the rise would not be insignificant and would amount to a major redistribution of the tax burden away from the motorist and would incite claims that it would both encourage people to drive more, and would be regressive against the lowest quartile.

Summary

If the move to BEVs is successful, revenue (duties/VAT/sales tax) from gasolene/petrol/diesel will disappear and will have to be replaced. There are alternatives, but all of them are unattractive.

The Payment for Charging Issues

The question of whether BEV drivers are being overcharged at public charging points, particularly at fast and rapid charging stations, has become a significant concern as the adoption of BEVs increases. Part of the impression of being overcharged lies in the direct comparison that the BEV driver can make between the cost of home charging and the prices being charged at public charging points.

BEV owners who park in their own garage or on their own driveway, and own their own home, have the opportunity to install a home BEV charging point and recharge at home, typically benefitting from lower electricity rates, especially if they can take advantage of off-peak tariffs.

I will use UK figures again as an example, but the comparative figures are similar around much of the world (with some notable exceptions).

To recharge at home requires the investment in the installation of home charging point equipment (about £1000–£1500 in UK/US$ 1250–$1900). Otherwise you are restricted to trickle recharge via a mains cable charger.

The cost per kWh for home charging can range from £0.10 to £0.20 per kWh, depending on the energy provider and time of use.

By comparison, public charging, especially at fast and rapid recharge points can range from £0.30 to over £0.70 pence per kWh.

The price difference is justified as being due to the investment costs of providing the infrastructure, operational expenses, and the 'reasonable' profit margins of the charging network operators. The wide price range lies in the different types of charging point.

A 'Fast' charging (7–22 kW) point typically charges around £0.30–£0.50 per kWh.

A 'Rapid' charging (50–150+ kW) point typically charges £0.60–£0.70 per kWh.

The high cost being claimed by the providers as being the additional costs of providing the rapid charging infrastructure, but is, at least to some extent, the opportunity cost for the convenience of faster charging times. It can lead to significant expenses for drivers relying on these chargers regularly.

However, setting up and maintaining fast-charging infrastructure is costly. This requires the installation of high-power equipment, grid connection upgrades, and ongoing maintenance, so a price differential is appropriate.

The electricity used at fast and rapid chargers is often procured at peak times, which is more expensive. This cost is of course passed on to the consumer.

Charging point operators are normally businesses, not charities and seek to recover their investment and generate and maximise their profitability. I say normally because some governments (e.g. China, Norway) provide subsidies to both installers and users to encourage BEV take up.

Although the cost per kilometre for BEVs charged at home is generally cheaper than for petrol or diesel vehicles, the gap narrows significantly when using public fast, and especially rapid, chargers. In some cases, the cost per mile for a BEV using rapid public charging can be comparable to, or even exceed, the cost per kilometre for a petrol or diesel vehicle.

The situation is adversely compounded because the BEV charging market is still evolving, and pricing can vary widely between different providers and locations. This lack of standardisation can lead to confusion and make it difficult for consumers to know if they are getting a fair deal.

Some charging networks offer subscription plans that can reduce the cost per kWh, but these may not be cost-effective for occasional users.

Howsoever the facts, many BEV drivers feel that the cost of public charging, especially at rapid chargers, is too high. This perception is compounded by the fact that they are paying significantly more than they would for home charging. Those who do not have the advantage of being able to recharge at home feel particularly disadvantaged.

While it is not universally accurate to say that BEV drivers are being 'ripped off', it is clear that public charging, especially at fast and rapid chargers, is significantly more expensive than home charging. Service stations on motorways/autoroutes are used to, and get away with, charging much more per litre (+10% to +20%) for petrol and diesel than petrol purchased at a supermarket, and see no reason to change their business model, indeed, because of 'range anxiety' see the opportunity to ramp up margins further.

There are calls for greater regulatory oversight to ensure fair pricing at public charging points, as the disparity in costs can be seen as a barrier to BEV adoption. The move to use BEVs is being driven by political pressures, not market pressures, so it is probably inappropriate to leave implementation solely to market forces.

These high costs can create a financial burden for those who rely heavily on public infrastructure, particularly if they have limited access to home charging (remember from above more than 40% of UK houses rely on on-street parking and have little or no opportunity to recharge at home). To date almost all BEV drivers are home-owners with garages or drives. As the market matures, it is hoped that increased competition, regulatory oversight, and technological advancements will help bring down these costs. The reality may very well be that the high cost of fast/rapid charging will deter the on-street parking driver from ever switching over to driving a BEV.

Just how Green Is a Current BEV?

The objective of BEVs are to eliminate the use of gasolene/petrol/diesel to power cars, and by so doing significantly reduce emissions that lead to global warming and pollution in general, but, as we have seen in the chapters above, their 'green' credentials are more complex when examined in detail. The environmental impact of BEVs can vary significantly depending on factors such as the fuel sources used for electricity (and hydrogen) generation, the materials and processes involved in BEV battery production, and the lifecycle of the vehicle.

As we discussed in Chap. 5 'Zero tailpipe emissions v lifetime emissions', if a BEV is charged using electricity generated from renewable sources (such as wind, solar, nuclear, or hydroelectric power), its operation can be almost carbon-neutral, making the BEV significantly greener than a vehicle that is powered from hydrocarbon sources.

But the reality (depending on the country it is recharging in) is that, unless you live in Scandinavia or France, currently 49%–70% of the electricity used to power BEVs during their working lifetime is not 'green', but carbon based. China may have over ten million BEVS, but at the present time, 66% of its electricity is generated from hydrocarbons, and China is not committed even to carbon neutrality (carbon neutrality, not zero use of carbons) until 2060. So, for the moment, in China, 10+ million BEVs, which each weigh a third more than an equivalent ICEV, so use more power to move, are belching out more carbon emissions somewhere in the country, than their ICEV equivalents, for two thirds of their working lives. The situation is similar in most countries (except Scandinavia where most electricity is hydropower generated, and France where most power is nuclear).

To compound the issue, while the general energy use in the manufacturing process is similar for BEVs and ICEVs (the ICEV engine block takes more energy to make than a BEV motor, but there are frequently multiple motors in a BEV), BEV battery production is a high energy consumption affair. As we have seen in the above chapters, the production of BEV batteries requires significant amounts of raw materials such as lithium, cobalt, nickel, and graphite. The extraction of these materials often involves significant greenhouse gas emissions. The process of manufacturing BEV batteries is highly energy-intensive. It is estimated that producing a single BEV battery can result in emissions equivalent to several tons of CO_2, depending on the size of the battery and the energy mix used in the manufacturing process. An analysis by the International Council on Clean Transportation (ICCT) calculated that an ICEV such as a Volkswagen Golf causes around 7.2 tonnes of carbon emissions during its production process. In comparison, a similar-sized BEV, such as a Volkswagen ID.3, causes around 9.2 tonnes—some 28% greater. As a result, the ICCT calculated that the BEV produced the equivalent of over 7000 miles of ICEV carbon emissions more than the ICEV to manufacture—the equivalent of about a year of driving.

BEVs have fewer moving parts than ICE vehicles, and this can lead to lower maintenance requirements. This can further reduce their environmental impact by reducing the demand for replacement parts and the associated manufacturing emissions. But the vehicle lifespan is determined by the life of the fitted battery. (Few would replace a battery that cost 35–50% of the cost of a new vehicle into an aged relic); and a BEV has a much greater risk of being an insurance write-off.

And as we have seen in previous chapters, when it comes to 'End-of-Life' and recycling, recycling BEV batteries is challenging and currently is not widely practised. The process is complex, expensive, and not all materials can be recovered efficiently. So, while BEVs are often touted for their sustainability, the materials used in their batteries and other components can be difficult to recycle, potentially leading to significant amounts of waste at the end of the vehicle's life.

So, in summary, at the moment, BEVs can claim justifiably that they are a little bit greener than ICEVs, but not by much. And they are far from the 'zero emissions' vehicles of manufacturer's claims. The situation will improve over time as energy production decarbonises and assuming that solutions to the battery recycling issues are found, but it will take more than a decade—perhaps in some countries three decades—before BEVs can claim to be a 'green' mode of transport.

Has Anyone in Government Bothered to Think this Through Properly?

When they are in power, politicians are supposed to operate to satisfy their goals through a combination of strategic planning, policy formulation, stakeholder engagement, and public communication. Politicians often base their goals on campaign promises (manifestos) and the 'mandates' to fulfil those manifestos given to them by voters. This includes addressing key issues highlighted during elections, such as the economy, healthcare, or climate change. And the encouragement to switch over to driving BEVs is a legitimate response to public concern for Government to take action to limit the causes of global warming.

Effective politicians use data and research to identify pressing issues and opportunities for intervention. This may involve analysing public opinion, economic indicators, and demographic trends to set realistic and relevant goals. Good politics is the art of achieving the possible. Bad politics is promising the unachievable. The objective being to ensure that the proposed policies are feasible and have the necessary support. Seeking to fulfil unachievable manifesto promises has led to public backlash with the consequences of a price to be paid by the politician who attempts to achieve the unachievable and the subsequent downfall of many a government around the world.

Politicians achieve their goals through a structured process that involves setting clear objectives, formulating policies, engaging with stakeholders, effectively communicating with the public, and monitoring outcomes. Achieving goals requires adequate resources, including funding, manpower, and infrastructure. The combination of strategic planning and public engagement is crucial for achieving sustainable political success and ensuring that policies meet the needs of the constituents.

The European Union (EU) has declared that it will ban the manufacture of new ICE vehicles from 2035 as part of its broader strategy to combat climate change and reduce greenhouse gas emissions; therefore worthy of a political statement. The EUs declared logic for the ban is closely tied to its commitments under the Paris Agreement, which aims to limit global warming to well below 2 ° C above pre-industrial levels. The EU calculates that reducing emissions from the transportation sector, which is a significant source of CO_2, is crucial to achieving these goals. The ban is part of the EU's broader 'European Green Deal', which aims to make Europe the first climate-neutral continent by 2050. The transportation sector is responsible for nearly 25% of the EU's total greenhouse gas emissions, and transitioning to zero-emission vehicles is seen as a critical step in reducing this impact.

But why a total ban? Other than a political gesture? The EU has argued that advances in battery technology, reductions in costs, and the increasing range of BEVs make the transition from ICE vehicles to BEVs feasible by 2035. (Feasible, 'desirable' and 'practical' are not the same thing: it is feasible to live if you only have one leg—that does not make it desirable). The anticipated improvements in EV infrastructure, such as charging networks, are also targeted to support this transition. That something like 40% do not have access to at-home charging and the cost and inconvenience of public point charging make running a BEV impractical for many of them, is not addressed.

The EU has highlighted that banning new ICE vehicles will contribute to improved air quality, especially in urban areas, leading to better public health outcomes. The shift from ICE vehicles to BEVs also supports the EU's goal of reducing dependency on oil imports, enhancing energy security, and stabilising energy prices.

In order to mitigate the economic impact on industries and regions heavily reliant on automotive manufacturing, the EU has established the 'Just Transition Fund'. This fund is designed to support workers and communities that will be affected by the transition, providing financial assistance and retraining programmes. The EU is investing in research and innovation to support the development of new technologies, including more efficient batteries and hydrogen fuel cells, to ensure that European automakers remain competitive in the global market.

Recognising the need for widespread and accessible charging infrastructure, the EU has set targets for the installation of public chargers across member states. This includes funding through the 'Connecting Europe Facility' and other EU programmes to ensure the necessary infrastructure is in place by 2035.

The transition to BEVs requires significant upgrades to the electricity grid to handle the increased demand. The EU is working with member states to plan and implement these upgrades, focusing on integrating renewable energy sources to power the grid sustainably.

But the EU is in an odd position as a 'government' because the actual means and steps taken to implement its decisions are reliant on, and taken by, the governments of its 27 member states.

Incentives for consumers to purchase BEVs, such as subsidies, tax breaks, and rebates, have been implemented by various member states with the support of EU policies. These incentives are intended to make BEVs more affordable and attractive to consumers, facilitating the transition away from ICE vehicles. But in many Member States, those benefits once given, have since been reduced or withdrawn.

The UK position is worse. To protect EU market access for vehicles, it manufactures it first replicated the EU regulation, and soon after its then gung-ho (overly enthusiastic) Prime Minister, in an 'anything you can do we can do better' bravado moment, declared 'if the EU can do this by 2035, UK can do this by 2030'. (Since reversed, then put back in place by the administration elected in 2024). But the UK is no longer a member of the EU so its companies and citizens do not have access to the incentives offered by the EU, and apart from some one-off deals with its car makers, the removal of subsidies to buy a new BEV, and the removal of annual vehicle tax exemption for BEVs, have largely left it to market forces to implement.

When the UK Government announced laws forcing carmakers to sell more electric vehicles (the ZEV Mandate)—a copy of similar legislation in the EU, except the EU laws allow PHEVs while the UKs law does not, UK ministers claimed that 'it would put Britain at the vanguard of the clean energy transition'. The reality is somewhat different.

The zero-emission vehicle (ZEV) mandate requires 22pc of new cars sold in the UK to be electric in 2024, rising annually to 80pc by 2030. Some clown in the government decided that this would give auto manufacturers the certainty they could ramp up production of new BEVs, while providing a steady stream of supply to the much bigger used car market. Manufacturers who do not comply face fines of £15,000 for every car that goes over their non-green quota.

Yet as the Government's drive to increase BEV ownership is stalling (sales from January to July 2024 rose by 10pc this year, that compares to an increase of about 18pc the previous year), it has left auto manufacturers worrying they could struggle to meet their ZEV mandate targets, with some resorting to withholding supplies of petrol cars in a bid to artificially push their percentage share of EV sales higher in order to avoid fines.

It has made more expensive BEVs an even harder sell, forcing manufacturers who want to meet their ZEV mandate targets to either discount heavily or cut the number of petrol cars they are selling, reducing their achievable sales and their profitability, even challenging their business viability. This aggravates the depreciation rates of all BEVs, and by the time they are reaching 5 years old they have now dropped to price party with equivalent ICEVs although they cost a third or more to purchase new.

And this problem upsets the functioning of the rest of the car supply chain, including the used market. This is because of the impact on fleet operators—such as car leasing companies and rental firms—who currently account for about 80% of BEV sales (source: Society of Motor Manufacturers and Traders). Their business model typically involves owning a car for a year and then reselling it. But BEV values are depreciating so quickly that it is putting many fleet operators under serious pressure.

In September 2024, Reuters reported that Italy's Industry Minister, Adolfo Urso said that the European Union should decide already next year whether to rethink regulation that effectively bans the sale of new petrol and diesel cars in the 27-nation bloc from 2035. Speaking on the sideline of TEHA business forum in Cernobbio, Adolfo Urso said an intermediate assessment on the progress to meet the new rules, currently due in 2026, should instead take place at the beginning of next year. 'I think this should be the first issue the new European Commission deals with, because companies and workers need clarity', said Urso.

Agreed EU law will require new cars sold after 2035 to have zero-carbon emissions, making it impossible to sell new fossil fuel-powered vehicles in the region.

The right-wing government of Giorgia Meloni believes the European Commission should allow member states to have more freedom in choosing the technology they prefer to meet agreed decarbonisation targets and supports a more gradual shift away from ICEV technology.

'The 2035 ban on new combustion engine cars is absurd and needs to be revised', Italy's Energy Minister Gilberto Pichetto Fratin said at the same conference.

Two years ago, the average BEV would be expected to depreciate by 40% over a 3-year period and retain 60% of its initial value, according to data from the British Vehicle Rental and Leasing Association (BVRLA). Today, they are depreciating by around 65% and retaining just 35% of their purchase value—and the decline is continuing to get worse.

This matters to the vast majority, who acquire their car on hire purchase or lease, because, along with rising interest rates, it is pushing up the cost of car finance deals, which typically see the consumer pay the cost of a car's depreciation over the set period. Leasing companies are now seeking to reduce their own exposure to that risk by asking consumers for bigger deposits or monthly payments. The BVRLA commented (source The Telegraph) 'We have seen double-digit percentage increases in lease rates in recent years, which has made leasing a new EV less attractive to businesses and consumers'.

The blanket ZEV target requirements are destroying the commercial viability of the automotive industry.

China, as we have seen above, has taken a different approach. It has not banned ICEVs, but set a goal that by 2035 all new vehicles should be 'New Energy Vehicles' (BEV, PHEV, Fuel Cell) and provided generous subsidies to auto manufacturers, infrastructure providers, and car purchasers to follow that path. As a result, last year, EVs (BEV/PHEV) accounted for approximately 24.7% of all car sales in China. By comparison, BEVs made up about 14.6% of the total car market in the EU. In the UK, BEVs constituted around 16.1% of new car sales. In the USA, BEVs represented about 7.2% of all car sales.

I leave the reader to assess which policy has been the most successful.

But with a zero-emissions target for electricity production of 2050, while it is easy to understand encouragement and incentives to move to BEVs beforehand, banning ICEVs, particularly PHEVs, some 15 years ahead of zero emission, will apply unnecessary pressure to the roll-out of public charging infrastructure, and may deter BEV take-up, for all the reasons given in Chap. 4. This is harder to understand, particularly when you consider the other current deterrents to BEV take-up discussed above.

Has Anyone in Government Outside China Considered How to Address the Circa 40% + who do not have Access to At-Home Charging, and the Consequences of not Doing so?

For the likelihood is that the consequence will be that a large proportion of them will hang on to ageing gas-guzzling ICEVs, thus setting back achievement of zero-carbon goals. Whereas allowing PHEVs would at least in part address the issue.

And for those governments who regulate to disrupt to achieve a political goal, however worthy, but then leave it largely to 'market forces' to implement—woe betides you at the next election!

Chapter 7
Ban Hydrocarbon Cars by 2035 (or Should It Be 2030?)

Abstract Chapter 7 compares the ZEV/Zero Tolerance strategies of Western countries with the Chinese 'New Energy Vehicles' strategy and concludes not only that the 'New Energy Vehicles' strategy is more efficient at reducing CO_2 emissions, but that in the period to 2040, maybe 2050, zero tolerance policies may add to the emissions problem because much of the energy used is not green.

Keywords Public charging points · Green electricity generation capacity · Lowest income quartile · On-street parking · New energy vehicles · Range anxiety · BEV only strategy

The short answer to this question is that to do either is counterproductive to the goal of reducing carbon emissions from the transportation sector.

We learned in Chap. 4 'The challenges of fuel supply–Domestic infrastructure', and using the UK as a typical example, that 'There are 40.5 million cars and light commercial vehicles in the UK. There are approximately 28 million homes in the UK. About 60% of homes in the UK have access to off-street parking, which includes both drives and garages. This translates to roughly 16.8 million homes with some form of private parking. It means that about 40% of homes, roughly 11.2 million homes, rely on on-street parking. That doesn't 1:1 equate with 40% of the 40.5 million light vehicle population because those with off-street parking are also more likely to own multiple vehicles.

Approximately 20% of UK homes are flats or apartments. This means there are around 5.6 million flats or apartments. Many of those apartments have shared off street parking but no access to a personal power supply. Many others have to rely on on-street parking. Conversely, many of the 5m+ light commercial vehicles are parked overnight on business premises.

It is difficult to obtain more precise figures, but probably true to say that in excess of ten million vehicles—**a quarter of the light vehicle park—do not have the means—nor the possibility of fitting the equipment—to recharge a BEV.** The vast majority of that 25% of the light vehicle sector will therefore not buy/lease/rent a BEV'.

B. Williams, *Electric Shocks: Are Drivers Rejecting the Electric Car?*,
https://doi.org/10.1007/978-3-031-98820-2_7

Until that problem is solved, and providing more public charging points may help, but will not solve the problem, as I surmised in the last chapter, about 25% of the population will hang on to ageing gas-guzzling ICEVs, thus setting back achievement of zero-carbon goals.

We learned in Chap. 3 'Source of human-generated CO_2 emissions' that cars and light vans are responsible for 14% of CO_2 emissions. 25% of that—*3.5% of all CO_2 emissions—can be seen as the probable CO_2 emissions consequence of not addressing the problem of drivers using on-street parking hanging on to their ageing ICEVs.*

On the issue of providing more public charging points, as we noted in previous chapters, building the green electricity generation capacity and distribution means to power 100% BEV population will likely not be possible by 2035. Rolling out the charging points by 2035 also presents a major challenge because these suffer from the same investment need and inconvenience factors of any public charging point, and unless subsidised result in an expensive price for the recharge, to be paid in most cases by the lowest income quartile householders that form much of the on-street parking community.

Pulling back from these counterproductive politically inspired bans and replacing them with 'New Energy Vehicles' targets, allowing PHEVs and encouraging the production of long EV range PHEVs, would at least in major part address the issue by addressing the 'range anxiety' and 'flat battery' concerns of the on-street parking community, and reduce their feeling of rip-off cost dependency on the local monopoly supplier charging point. It would also eliminate 'range anxiety' among the rest of the driving community, leading to a much faster 'norm' of driving using electric power for most of the time.

Recent studies have shown that the percentage of ICEV drivers who are hesitant or unwilling to switch to electric vehicles due to 'range anxiety' varies across different studies and regions, but it remains significant and generally ranges between 20% and 40% (Deloitte Global Automotive Consumer Study (2022): 20%–30%; McKinsey and Company (2021): 30%; AAA Consumer Survey (2020): 40%; Ernst and Young Study (2021): 20%–25%).

Range anxiety remains one of the top psychological barriers to BEV adoption, along with concerns about cost and charging times. So this change would not only affect the probability of the on-street parking community switching to NEVs and driving electrically powered most of the time, but significantly encourage the reticent off-street parking community to switch over as well.

If cars and light vans are responsible for 14% of all CO_2 emissions, if even 20% of the residual 75% of BEV reticent off-street parkers can by this measure be persuaded to switch to NEVs, and drive using electricity 90% of the time, this would save 2% of all CO_2 emissions—at 30%, 2.8%—so say 2.5% overall reduction of CO_2 emissions—between the on-street and off-street parkers, reducing all CO_2 emissions by some 6% (a quarter of transportation CO_2 emissions) better than the current BEV only policies.

These calculations are very crude but demonstrate clearly the benefits of a NEV strategy over a BEV only strategy (and are supported by the rate of NEV adoption in China).

Chapter 8
The Current BEV Is Proving to Be ‘Not Fit for Purpose’

Abstract Chapter 8 opines that while there are hopeful signs for the future, the current BEV is proving to be ‘not fit for purpose’, and the signs are there that the wider driving population are not adopting the same acceptance as the largely affluent middle-class early adopters. The chapter discusses—What does ‘fit for purpose’ for a family/personal vehicle require?—Movement from A to B on demand—ease of refuelling—availability of refuelling (home and away)—time to refuel—affordability of refuelling, affordability of vehicle purchase price.

Keywords Middle-class early adopters · Falling sales rates · Range anxiety · Refuelling availability · Time-to-refuel · Lowest quartile income groups · Price differential

The Signs Are There that the Wider Driving Population Are Not Adopting the Same Acceptance as the Largely Affluent Middle-Class Early Adopters

The earliest adopters of BEVs were almost exclusively the climate-concerned middle/upper middle class. They were affluent—they had to be because of the significantly higher purchase/lease costs of a BEV. They could also easily afford the cost of installing a home charging station. Environmentally aware company cars followed, largely to the same social groupings. As we have seen in Chaps. 3 and 5, in the period 2012–2018 sales growth was modest, but encouraged by governments and new more attractive models, BEV sales growth has boomed in the period 2019–2022 and levelled off in 2023.

The concentration of BEV adoption in urban and affluent areas is supported by several studies, which highlight how factors such as income, infrastructure availability, and lifestyle, contribute to this trend. Numerous studies have shown that BEV owners are more likely to be from higher income households. For example, a report by the International Council on Clean Transportation (ICCT) found that BEV

B. Williams, *Electric Shocks: Are Drivers Rejecting the Electric Car?*,
https://doi.org/10.1007/978-3-031-98820-2_8

adoption is strongly correlated with higher income levels. Affluent consumers are more likely to be able to afford the higher upfront cost of BEVs.

California, a leading state in BEV adoption, provides a clear example of this trend. The California Air Resources Board (CARB) reports that BEV adoption is concentrated in wealthy, urban, areas such as the San Francisco Bay Area, Los Angeles, and Silicon Valley. These regions have high average incomes, dense charging networks, and a strong environmental culture, all of which contribute to higher BEV adoption rates.

Similar trends are observed in Europe. In the UK, BEV registrations are significantly higher in London and other affluent urban areas compared to rural regions. A study by the European Environment Agency (EEA) also shows that countries with higher urbanisation levels and GDP per capita, such as Norway, the Netherlands, and Germany, have higher BEV adoption rates.

But there are several signs that suggest the broader driving population, particularly those outside the affluent middle-class early adopters, may not be adopting BEVs at the same rate. Some of these signs include falling sales rates in certain regions, persistent concerns about affordability, range anxiety, and a lack of widespread infrastructure.

In some regions, especially where incentives have been reduced or where early adopters have already made the switch, BEV sales growth has begun to stagnate or even decline. For instance, in 2023, there were reports of slowing BEV sales growth in markets such as the UK, Germany, and other parts of Europe, where high upfront costs and concerns about charging infrastructure, and the removal of subsidies and tax breaks remain significant barriers.

A further factor impacting the growth of BEV sales has been that as government subsidies and incentives for BEVs are scaled back or phased out, there has been a noticeable impact on sales. Middle and lower income consumers, who are more price-sensitive, are less likely to adopt BEVs without substantial financial incentives.

According to 'The Register' and other sources, consulting firm J.D. Power lowered its sales forecast for EV growth from 12% this year (2024) to just 9%. According to the European Automobile Manufacturers' Association (ACEA), BEVs accounted for 12% of the EU car market in July, down from 13.5% the previous year. New registrations of BEVs dropped by 10.8%, with shrinkage in countries such as Germany (down 36.8%) offsetting gains elsewhere.

Ola Källenius, CEO of Mercedes, in an interview with Auto Car magazine in August 2024 confirmed that Mercedes-Benz is set to phase out its EQS electric model just 3 years after its debut. The decision comes as the EQ series has fallen short of expectations. This move reflects a shift in Mercedes' strategy as the growth of its electric vehicle lineup has been slower than anticipated.

In September 2024 Volvo, which has traditionally flaunted its environmental credentials, announced that it too has abandoned its target to produce only fully electric cars by 2030, saying it now expects it will also be selling some hybrid vehicles by that date. The car maker blamed changing market conditions for its decision to give up a target it had announced only 3 years ago. Volvo now expects at least 90% of its output to be made up of both electric cars and plug-in hybrids by 2030. 'It is clear that the transition to electrification will not be linear, and customers and markets are moving at different speeds'.

In the USA, without these draconian penalties, Ford chief financial officer John Lawler, noting that the company is responding to market demand, told reporters recently that Ford would be 'pivoting' away from its existing electric future and instead expand its other platforms. 'What we've learned is that customers want choice, and so we're providing that choice, with a full lineup of EVs, hybrid, electric, gas and diesel products', said Lawler. 'The market is changing globally, with the Chinese at overcapacity, and so it's all going to come down to being nimble, having a low cost and multiple low-cost platforms, and being able to adjust as we continue to develop the transition of this industry', said Lawler. He said that softening EV sales were due to auto customers' 'shifting requirements', and stated that Ford had to go where the market was.

'It's coming back to understanding the customer, understanding how this is going to transition over time', Lawler said in his media briefing. 'It's about providing them those choices that meet their duty cycles and their needs, and that is giving them the options between full battery electric vehicles, hybrid technologies'.

Lawler said that Ford would not launch any EVs in the future unless they can be profitable within 12 months.

'We are launching multiple electric vehicles in Europe this year', Ford said in a statement, referring to the EU-only Ford Explorer EV and the Capri built on the same platform borrowed from rival VW's ID.4. 'We are adjusting the company's North America vehicle roadmap to offer a range of electrification options designed to speed customer adoption, including lower prices and longer ranges'. (source: Wired).

In terms of total electric vehicle production in the UK, electrified (BEV, PHEV, and HEV) vehicle manufacturing dropped 18.6% year-on-year in July. (Overall UK car production declined 14.4%.) (source: The Register).

BEVs typically have a higher purchase price compared to ICEVs, even when considering total life cost of ownership. This makes them less accessible to lower income groups. The majority of BEV sales to date have been concentrated among wealthier consumers who can afford the premium price, leaving a gap in adoption among the wider population. And while the number of BEV models on the market has increased, many of these are still positioned as premium vehicles. Affordable BEVs are less common, which limits adoption among consumers who have to prioritise budget over advanced features, performance, and environmental concerns.

Even though newer BEVs offer significantly improved ranges, as we have seen in Chap. 4, and in the previous chapter, 'range anxiety' remains a common concern for 20%–40% of drivers, and particularly among consumers who cannot afford the latest, long-range models. For many potential buyers, especially those living in rural or less densely populated areas, the fear of running out of charge with no nearby recharging stations is a major deterrent. In areas where charging stations are scarce or unreliable, potential buyers may opt to stick with ICEVs/PHEVs which they perceive as more convenient for long-distance travel and daily use.

BEVs are perceived to have uncertain resale values, partly due to rapid advancements in battery technology (fear of obsolescence) and concerns about battery degradation. This uncertainty can deter buyers, particularly those who are not financially secure enough to take on the risk of the potentially higher depreciation. The secondary market for BEVs is still developing, and many potential buyers are cautious about purchasing used BEVs due to concerns about battery life and replacement costs. The more competitive pricing and higher specifications—particularly range, of new model BEVs, depress second hand prices further.

And it must not be forgotten that many drivers remain attached to ICEVs due to familiarity, the driving experience, the perceived reliability of internal combustion engines, and the extensive network of refuelling stations. This attachment is stronger among those who prioritise convenience and experience and are less convinced by the claimed environmental benefits of BEVs.

Importantly, lower income households, which make up a large portion of the driving population, are less likely to adopt BEVs due to financial constraints. The upfront cost, combined with concerns about the cost of commercial charging points, insurance, long-term maintenance and battery replacement, makes a BEV a less attractive option. As we have seen, the less well-off are also less likely to be able to recharge at home on an economy rate overnight tariff and would have to rely on more expensive public charging points—an important deterrent to BEV ownership.

The charging infrastructure is more developed in urban areas, making it easier for city dwellers to adopt BEVs. For instance, data from the US Department of Energy shows that public charging stations are disproportionately located in urban centres. This urban concentration of infrastructure directly influences where BEVs are more popular. Research by the International Energy Agency (IEA) highlights that the availability of charging stations is a critical factor in BEV adoption.

While early adopters, primarily affluent middle-class consumers, have driven initial BEV adoption, the broader driving population remains hesitant. This is reflected in slowing sales growth, ongoing affordability issues, infrastructure concerns, and cultural resistance. For BEVs to achieve mass-market penetration, these barriers will need to be addressed through more affordable models, better charging infrastructure, and targeted incentives that appeal to a wider range of consumers. But even then many of the challenges appear insurmountable.

Back-to-Basics: What Does 'Fit for Purpose' for a Family/ Personal Vehicle Require?

This section is not designed to address all the issues regarding a 'fit-for-purpose' vehicle in general, but to identify the aspects of a BEV that would make an average head of a family decide that a BEV was not 'fit-for-purpose' for his/her families' needs.

The phrase 'fit-for-purpose' refers of course to something being suitable and capable of fulfilling the requirements for which it was intended. When evaluating whether a family or personal vehicle is 'fit-for-purpose', some general factors will apply in all cases. Factors such as: general safety features, compliance to safety and emissions requirements and regulations, technology, these days connectivity, info-tainment, and availability of options and accessories that can adapt the vehicle to specific needs (e.g. roof racks, child seats, bike racks), must all be taken into account, depending on the particular needs of the user, but are not affected by the power source of the vehicle and are the same considerations for ICEVs, BEVs, and PHEVs, alike.

Despite the issues with Tesla reliability reported in Chap. 3 above *(2010–2020 Enter Nissan, Tesla, Volvo and the big established auto-manufacturers)*, there is now ample evidence that the mechanical reliability and durability of BEVs is at least as good as that of ICEVs, possibly better. As already discussed in Chap. 4, tyre wear and brake wear are heavier in BEVs, the motors driving BEVs are simpler and easier to maintain.

The seating capacity of BEVs is as good, if not better, than its ICEV equivalent. BEVs tend to be larger in order to accommodate longer range batteries, therefore roomier, but cargo space and accessibility is sometimes impacted detrimentally by the space taken up by the battery pack.

Ride comfort and performance are two of the strengths of BEVs over ICEVs. Because of the position and weight of the battery pack, most BEVs have a low centre of gravity and handle well with rapid acceleration available though heavier braking is required to stop the additional weight.

Purchase Price: The vehicle has to be within the buyer's budget—cash outlay at purchase and ongoing monthly payments. Here we have our first significant difference and valid reason for not choosing a BEV.

In most countries, BEVs have significantly higher initial purchase prices than comparable ICE vehicles, generally around a third more expensive (largely due to the cost of the battery). Cheaper models are becoming available, but for a car with like-for-like features, expect to pay significantly more for a BEV. For example,

US and UK prices of the Hyundai Kona, which is available as an ICEV, BEV, and PHEV hybrid:

1. ICEV (Hyundai Kona SE): US$ 22,000–US$ 24,000 (UK£ 26,000).
2. BEV (Hyundai Kona Electric SE): US$ 33,500–US$ 37,000 (UK£ 32,000–£ 39,000).
3. PHEV (Hyundai Kona Hybrid): US$ 27,000–US$29,000 (UK£ 30,000).

The ICEV version is the least expensive, followed by the PHEV, and then the BEV, which is the most expensive. The BEV version is about $11,500–$13,000 more expensive than the ICEV, while the PHEV (or hybrid) is about $5000–$7000 more expensive than the ICEV.

One of the principal reasons for this difference in price being that the fuel tank for the Kona costs $50, the battery for the BEV Kona $12,000–$15,000. The PHEV has a much smaller battery than the BEV, so is less costly than the BEV despite the fact that it has to provide both means of motive power.

This is an important factor in the choice of whether or not to buy a new or used BEV, especially for the 'average' income family. The average salary in the USA is $60,000, in Germany $53,000, in the UK, and France $43,000. So a price difference of $11,000–$13,000 onto the cost of your replacement car represents 20–30% of total income before tax, and 30–40% net post tax annual income of an average single-salaried family; more for the lower quartile income groups. Even for a two income family, 20% of net income— not 20%–40% for the acquisition of a new vehicle, but just to opt for a BEV over an ICEV—and this presents a huge barrier to the purchase of a BEV.

While the UK (as an example) inflation rate has risen 26% since 2019, new car prices have increased by 40% since 2019, Auto Trader has reported, driven, they reported, by both higher price tags of BEVs as well as decisions by brands including Ford to stop selling lower priced models such as the best-selling Fiesta (source Auto Trader). Five years ago about 21% of new cars available were priced below £20,000 but only 4% are today (2024).

The cost of replacing the family vehicle is already a big financial hurdle for most average income families. Paying more for a BEV is something that most just cannot afford.

The option of buying second hand is also unattractive because even Hyundai, well known for its long warranties, will only provide an 8-year warranty on its BEV batteries, many offer less. So if you buy a 3–4-year-old BEV you risk being faced with an unaffordable bill for $13,500–$17,000 to replace the battery after a very few years.

These are issues which affected the affluent middle class early adopters far less, but are a brutally harsh barrier to the average family opting to buy a BEV.

Depreciation: The value of a car falls over its lifetime, and this depreciation in value is the size of most of the cost that you will have to fork out to replace your car (unless you are upgrading or downgrading), but as well as that, as we saw in Chap. 7, 'Ban hydrocarbon cars by 2035 (or should it be 2030?)' above, it affects the whole balance of the car sector, and well-intentioned government initiatives exacerbate the problem.

In the UK, the zero-emission vehicle (ZEV) mandate has further exacerbated BEV depreciation rates. (see Chap. 7 above).

It has made more expensive BEVs an even harder sell, forcing manufacturers who want to meet their ZEV mandate targets to either discount heavily or cut the number of petrol cars they are selling. This aggravates the depreciation rates of all BEVs, and by the time they are reaching 5 years old they have now dropped to price parity with equivalent ICEVs although they cost a third or more to purchase new.

As we saw in Chap. 7, 2 years ago, the average BEV would be expected to depreciate by 40% over a 3-year period and retain 60% of its initial value. Today, they are depreciating by around 65% and retaining just 35% of their purchase value—and the decline is continuing to get worse.

This matters to the vast majority, who acquire their car on hire purchase or lease, because, along with rising interest rates, it is pushing up the cost of car finance deals, which typically see the consumer pay the cost of a car's depreciation over the set period. Leasing/Finance companies are now seeking to reduce their own exposure to that risk by asking consumers for bigger deposits or monthly payments. The BVRLA commented (source The Telegraph) 'We have seen double-digit percentage increases in lease rates in recent years, which has made leasing a new EV less attractive to businesses and consumers'.

A further barrier to switching to a BEV.

Insurance Costs: This is another disincentive area for BEV ownership. As we learned in Chap. 6 'The problem of accident and insurance', BEVs introduce unique challenges for insurers. Higher purchase costs, higher repair costs, battery-related risks, and uncertainties around residual value and data availability all contribute to more expensive premiums .... At the moment, on like-for-like market value, BEV insurance premiums are likely to be up to 20% more expensive than their ICEV equivalent. But as the market cost of the BEV is about a third more than its ICE equivalent, the true difference that hits the driver's pocket is likely to be greater.

Running Costs: As we saw way back in Chap. 1 (The USP of the BEV) the driver of the ICEV pays fuel duty and VAT or sales tax on every litre/gallon of fuel purchased—lots of it (over half of what you pay at the pump is tax in many countries). Green electricity is claimed to be now less expensive than that generated from hydrocarbons and bears only a reduced rate of VAT or no tax at all in some countries. Even better is available at off-peak tariffs. So, for those who recharge their cars at home, off peak, Chap. 1 calculated 18 kWh per 100 km at an average of £0.10 per kWh = £1.80 per 100 km/£0.018 per km/£0.03 per mile (as opposed to 7 litres per 100 km at an average of £1.55 = £10.85 per 100 km/£0.109 per km/£0.175 per mile for an equivalent ICEV).

These are rough comparisons of course. Petrol prices vary daily, electricity prices are revised quarterly and tariffs vary, mpg and kWh usage vary with the route and driver. But in rough terms, fuelling a BEV off-peak costs one sixth of the cost of fuelling an ICEV.

For the average family car, driving 12,000 km per year/1000 km per month, that is an outlay of £18 per month/£216 per year as opposed to £109 per month/£1308 per year to fuel an ICEV.

That is so long as you can recharge at best off-peak rates.

If you have to recharge at peak times, the rate per kWh varies from £0.28 to £0.35 per kWh—say £0.30 for our comparison. So the comparison is 18 kWh per 100 km at an average of £0.30 per kWh = £5.40 per 100 km/£0.054 per km/£0.09 per mile (as opposed to 7 litres per 100 km at an average of £1.55 = £10.85 per 100 km/£0.109 per km/£0.175 per mile for an equivalent ICEV).

For the average family car, driving 12,000 km per year/1000 km per month, that is an outlay of £54 per month/£648 per year as opposed to £109 per month/£1308 per year to fuel an ICEV. About half the price of fuelling the ICEV.

But if you have to recharge using a public rapid charging point and remember that 40% of UK homes rely on on-street parking so have no opportunity to recharge at home, the rate per kWh varies from £0.50 to £0.70 per kWh—say £0.60 for our comparison. So the comparison is 18 kWh per 100 km at an average of £0.60 per kWh = £10.80 per 100 km/£0.108 per km/£0.174 per mile (as opposed to 7 litres per 100 km at an average of £1.55 = £10.85 per 100 km/£0.109 per km/£0.175 per mile for an equivalent ICEV).

For the average family car, driving 12,000 km per year/1000 km per month, that is an outlay of £108 per month/£1296 per year as opposed to £109 per month/£1308 per year to fuel an ICEV. About the same as the price of fuelling the ICEV—so offering no savings.

And if you are unlucky enough to have to pay the premium rate of £0.70 at the rapid charging station the comparison is 18 kWh per 100 km at an average of £0.70 per kWh = £12.60 per 100 km/£0.126 per km/£0.203 per mile (as opposed to 7 litres per 100 km at an average of £1.55 = £10.85 per 100 km/£0.109 per km/£0.175 per mile for an equivalent ICEV).

For the average family car, driving 12,000 km per year/1000 km per month, that is an outlay of £126 per month/£1512 per year as opposed to £109 per month/£1308 per year to fuel an ICEV. 15% above the price of fuelling the ICEV.

Depending on your personal circumstances, fuelling the BEV instead of an ICEV may be an advantage or a disadvantage. Probably in most cases, an advantage, but not so much as the salesman's repartee in Chap. 1 would have you believe.

Versatility and Adaptability: In order to be 'fit-for-purpose' a vehicle needs to be versatile enough to handle various types of driving (e.g. city, highway, rural) and activities (e.g. daily commuting, family road trips for holidays, excursions, visits to friends and family). These are also very important aspects that are addressed in the remaining subchapters of this chapter.

Movement from A to B on Demand

As stated in Chap. 1 (Social engineering, causes and consequences), those who live in cities can usually commute to a fixed work location, or locations, by public transport, cycling or walking, but many need a car to visit relatives and friends, do the school run and ferry children around their busy social calendars. Others need their car in order to access their workplace, or carry out their work: carers, maintenance workers, social workers, doctors, professional advisors, some delivery drivers, private car hire and many taxi drivers, etc. And then there are holidays, excursions, and out-of-town shopping trips. So, cars remain an essential requirement for most citizens. Many, indeed most, workplaces have also located themselves in the suburbs to avoid city centre congestion issues, and so one (or more) car(s) are essential for most families.

Driving your own car offers many advantages over using public transport: you are in control of your own personal space. In your car, you have full control over your environment, including temperature, music, and seating, with no crowding or (assuming you can control your family) noise from other passengers. Driving allows you to carry more luggage, groceries, or other large items that would be cumbersome on public transport. Families with young children, pets, or those with mobility issues can find it easier to travel with the added space and convenience of a private car. You can travel to remote destinations, rural areas, or take spontaneous road trips that may not be accessible via public transport.

The car allows you to travel directly from your home to your destination without transfers, waiting times, or walking to/from stations. You can choose your route, avoid congested areas, and take detours if necessary and you can take the most direct route, reducing travel time compared to indirect public transport routes. Except for very long journeys where a plane or train might be quicker, driving your car is usually the fastest option.

While all these aspects are significant, the three most important key advantages of travelling by your own car are, however, usually cited as:

1. It provides a single door-to-door service from A to B.
2. 24/7 Availability.
3. Faster travel.

Except

(A) Where you have had to park up at an inconvenient public access point to recharge the battery.
(B) You cannot drive the car because the battery needs charging.
(C) You cannot drive the car because the battery needs charging and there is no charging access point available at the time you need it.
(D) You cannot drive the car because the battery needs charging, and there is no charging access point available in the locality at all.

BEV drivers who recharge at home are usually only affected by B) through bad planning—except when they are travelling far from home, but this may affect on-street parkers often.

None of these issues affect ICEV or PHEV drivers.

So, these BEV-specific issues are a deterrent to BEV driving, especially for the on-street parked vehicles.

Ease of Refuelling

There are notable differences when comparing the ease of refuelling a BEV with refuelling an ICEV or FCEV. They come down to issues of convenience, time, availability of infrastructure, and overall user experience.

Let us start with refuelling an ICEV, with which we are all familiar.

Firstly, you have to drive to a service station. As there are probably around 1,500,000+ service stations in the world (each with 2–12 pumps—say 9–10 million pump points) you are not far from one and will almost certainly pass one or more on your journey (and if you are travelling in a remote region, you can always take a jerrycan of extra fuel with you).

It typically takes 3–5 min to fill a petrol or diesel tank at a fuel station, regardless of the car's fuel level. After refuelling, and paying, you can continue your journey immediately with a full tank.

If you are one of the lucky 60% who can recharge your BEV at home, and your journey is well within your BEVs achievable (not claimed) range, as it makes economic sense to recharge overnight, you simply have to plug in and activate the timer before you go to bed and you are topped up, ready-to-go, every morning. No trip to find a charging point needed.

But if you have a BEV that you park on-street, or are on a long journey, or had not topped up before you set off, you will have to find a public charging point. There are now calculated to be about 3,500,000 EV recharge points in the world (about a third as many as there are ICEV pump points) but more than half of those are in China, and whereas an ICEV pump point is tied up for 5 min per refill, an EV recharge point is typically tied up for 30–60 min per recharge (longer of course for slow overnight recharge points).

Recently, some automotive manufacturers have 'demonstrated' BEVs which can recharge in 5–8 min, but these largely require solid-state batteries (see above) combined with a new generation of very high power chargers. The infrastructure for which does not yet exist and will be impractical in many places because of the very high power drain on the local grid. But first you have to find a public EV recharge point that is available.

Availability of Refuelling (Home and Away)

In Chap. 4 'The challenges of fuel supply (domestic infrastructure)', we considered the issues of access to home charging points and concluded that it is difficult to obtain more precise figures, but probably true to say that in excess of 10 million vehicles—a quarter of the light vehicle park—do not have the means—nor the possibility of fitting the equipment—to recharge a BEV.

As stated in the last section, there are about 3,500,000 EV public recharge points in the world (about a third as many as there are ICEV pump points). These include both slow chargers (AC) and fast chargers (DC).

Of these three and a half million, only about 10,000 are 'Ultra Rapid' charging points and 55,000 are 'Rapid' charging points. (See the next section for more information about these categories.)

And remember also from the last section that whereas an ICEV pump point is tied up for 5 min per refill, an EV recharge point is typically tied up for 30–60 min per recharge (longer of course for slow overnight recharge points).

Consistent with the Chinese policy that I have explained in previous chapters, China is far and away the global leader in EV charging infrastructure. As of 2023,

China has about 1.8 million to 2 million public EV charging stations. Indeed, China accounts for more than half of the world's public EV charging points.

Europe has around 500,000 to 600,000 public EV charging points, with major contributions from countries such as the Netherlands, Germany, France, and the UK.

The USA has approximately 150,000 to 170,000 public charging points, with major charging networks such as Tesla Supercharger, Electrify America, and ChargePoint.

Canada has around 20,000 public EV recharge points; Japan, approximately 30,000 to 40,000 and Australia around 4000 to 5000.

Remember too, from the last section, that whereas an ICEV pump point is tied up for 5 min per refill, an EV recharge point is typically tied up for 30–60 min per recharge (longer of course for slow overnight recharge points).

As of 2023, the total global car population is estimated to be approximately 1.5 billion vehicles. This includes both passenger cars and light commercial vehicles.

The global number of BEVs is estimated at around 26 million as of 2023, so BEVs make up about 1.7% of the total global car population.

So there is about 1 public charging point (of all types) for every 7–8 BEVS, but only 1 ultra-rapid/rapid charging point per every 500 BEVs, and one ultra-rapid charging point per every 2600 BEVs—and that is with BEVs being just 1.7% of the car population.

So first you have to find a public EV recharge point that is available.

Time to Refuel

Charging a BEV can take anywhere from 30 min to several hours, depending on the type of charger used:

- Level 3+/4 Ultra-Rapid Charger (DC 150 kW–350 kW) typically takes 15–30 min to recharge to 80%.
- Level 3 Rapid Charger (DC 50 kW–150 kW) typically takes 20–60 min to recharge to 80%.
- Level 2 Fast Charger (AC 7 kW–22 kW) typically takes 4–12 h to recharge to 80%.
- Level 1 Charger (AC3 kW–7 kW) typically takes 8–20 h to recharge to 80%.
- Home charger (AC 7 Kw) typically takes 6–12 h to recharge to 80%.

Whatever charger you access, except the home charger overnight, it is an inconvenience, possibly a serious inconvenience. The shortage of ultra-rapid recharge points mean that it is likely to be a significant inconvenience.

Affordability of Refuelling

Issues concerning the cost of refuelling have already been discussed in Chap. 6 'The payment for charging issues'. This section deals specifically with affordability issues.

In Chap. 6, we concluded that the affluent who are able to recharge their BEVs at home are sitting pretty and able to derive operating cost benefits from driving their BEV. But for those who rely heavily on public infrastructure charging points, particularly if they have limited access to home charging (remember from above more than 40% of UK houses rely on on-street parking and have little or no opportunity to recharge at home), the cost burden of public charging point charges (outlined in Chap. 6) falls most heavily on those who use on-street parking, and particularly those who have to use rapid or ultra-rapid recharge points or the recharge point of a local monopoly supplier.

Who are the on-street parking community? Well we know that they do not have a garage or drive, and so, with a few exceptions in fashionable or historic areas, are people who live in terraced houses with little or no front garden, in flats over shops, apartments and flats, houses divided into bedsits, etc.

We can generalise that, on the whole:

Those who live in *terraced houses* have historically been working class families because historically, terraced homes were built for industrial workers, and today, they are still common in working-class neighbourhoods, especially in older urban areas. In some areas, older people who have lived in these homes for many years may continue to reside in terraced housing, particularly in communities where they have deep roots.

More recently, built terraced housing is likely to be 'starter' homes or social housing (because they are cheaper to build) where the occupants are most likely to be in the lowest income quartile.

But because these homes are often more affordable than detached or semi-detached homes it makes them attractive to those with lower or middle incomes. In urban areas, terraced houses can attract young professionals looking for affordable housing with more living space than apartments, especially in gentrifying neighbourhoods. In capital cities and large cities however, the price of housing means that you can find almost any social group in them, especially where people with two homes who have their main home out of town.

Flats above shops are often more affordable and are located in busy, central areas. Some people who own or work in the businesses below might live in the flats above the shops as a convenience option, or for economy. Younger people and students may live in such accommodation for its proximity to work, university, or nightlife. Flats above shops tend to be less expensive, attracting lower income tenants who are seeking affordable, central locations.

Many young professionals live in *apartments*, especially in cities, where they are convenient for work and social life. In large cities, the middle class often live in apartment buildings, especially in newly developed or luxury apartment complexes.

Many retirees choose apartments for ease of maintenance, security, and access to amenities. Apartments tend to suit single people or couples who do not need large living spaces.

Apartments and flats in more affordable urban areas are often home to immigrant families who need affordable housing and easy access to work and public transportation.

Bedsits are common among students and younger workers due to their affordability and the transient nature of their housing needs. People with low-wage jobs may rent bedsits or rooms in shared houses because it is often one of the cheapest housing options in expensive cities. Often, recent immigrants live in bedsits or shared accommodation as an affordable way to establish themselves in a new country. People on benefits or in temporary housing, in some cases, those on housing assistance or facing temporary financial hardship may live in bedsits or shared housing.

These housing types are located in urban areas or densely populated suburbs, where space is at a premium and housing is more affordable. They attract:

- Low- to middle-income groups seeking affordable living close to employment, education, or urban amenities.
- People in transition (e.g. students, young professionals, immigrants).
- Individuals and families who prioritise affordability and convenience over space and privacy.

While some of these occupiers will be affluent, the overriding majority will be in the lowest quartile income groups.

Because public access charging points have to recoup their investment, infrastructure, and operating costs and make a return for their investors, and unless subsidised, they result in an expensive price for the recharge, to be paid in most cases by the lowest income quartile householders that form much of the on-street parking community.

This therefore creates a punitively regressive charging system—a system where the costs or fees disproportionately affect those with lower incomes or fewer resources. It means that individuals or groups with lower financial capabilities and capacity end up paying a larger percentage of their income compared to wealthier individuals, making the fees more punitive and onerous for them. To put it bluntly, it is basically unfair and will prevent take-up of BEV usage.

Affordability of Vehicle Purchase Price

Following on from the logic of the previous subchapter, together with Chap. 6 'The problem of accident and insurance' where we introduced the fact that insurance premiums are affected by the fact that new BEV purchase costs are, on average, a third more expensive than their ICEV equivalent, and Chap. 8 'Back-to-basics—What does 'fit for purpose' for a family/personal vehicle require?: Purchase Price' where we see actual like for like compared purchase costs, it is pertinent to consider who buys, and who will be expected to buy a second-hand BEV, its affordability issues, and the consequences of them opting not to buy a BEV.

Let us consider how the used vehicle market works.

The used car market operates as a system in which used cars are bought and sold by individuals, dealerships, and sometimes manufacturers. It plays an essential role in making vehicles accessible to a broader range of buyers who cannot afford, or choose not to purchase or lease, a new car.

This second-hand market largely depends on the flow of new cars into the market. When consumers buy new cars, they usually trade in or sell their old vehicles, feeding the used car market. Similarly, when vehicle leases expire (typically after 2–3 years), leased cars are returned to dealerships, refurbished, and then resold as used cars. Also a major player in the first level of the used car market are car rental companies who sell vehicles from their fleets when they reach a certain mileage or calendar age.

Because a car has been used and incurred some wear-and-tear it usually sells for less than its price when new, or last sold. This drop in price is known as 'depreciation'. The amount that a vehicle depreciates by is determined in part by its calendar age, in part by its condition, in part by the number of miles/km it has been driven and in a large part by the demand for that make/model vehicle.

Many people, especially those with lower incomes, opt for used cars because they are cheaper than new cars and to avoid the steep depreciation of new vehicles. This leads to a trickle-down effect in the second-hand car market as cars move from higher income buyers to lower income buyers over time as they age and depreciate.

Wealthier buyers often purchase new cars. These cars depreciate rapidly in the first few years, but the owners often trade them in or sell them after a few years, typically for a newer model. Once these cars are resold (typically at 2–4 years old), middle-income buyers can afford them at significantly reduced prices. These lightly used cars still have good reliability but cost much less than when new. As these cars continue to age and are sold again, they become accessible to lower income buyers. By the time a car is 7–10 years old, it is often much cheaper, though with higher mileage and more wear and tear. It will likely pass through multiple owners before reaching the most budget-constrained buyers.

Eventually, cars that have gone through several ownership cycles may reach the end of their useful life for most buyers. These vehicles are often sold cheaply for parts, scrapped, or exported to developing countries, where they may continue to be used in markets with fewer regulations around vehicle age and emissions.

The used car market enables a wide range of people to own cars, especially those who would not be able to afford a new one. This democratises car ownership. Having access to a car can increase mobility for lower income individuals, providing more employment opportunities and access to resources, which may lead to improved economic conditions and lifestyle.

So if the marketplace reacts normally to BEVs being resold, and when new they were priced a third dearer than their ICEV equivalent, their initial depreciation will be greater but they will retain their price differential. That is they will remain the more expensive option as they hit a more price sensitive part of the market—which will deter take-up.

However, in the chapters of this work we have elaborated many age-related disadvantages of BEVs and the evidence is accruing that this is decreasing demand for second hand BEVs which is increasing their rate of depreciation, i.e. lowering their price (and current evidence is that by 3–4 years old the price of a BEV and an equivalent ICEV are about the same).

In a perfect market, the depreciation increases/market price falls until a buyer is found. But, as the chapters in this work have described, many of the prime obstacles to BEV ownership are not purely economic. So what happens if many used BEVs remain unsold?

The sale to developing countries is unlikely in any bulk because developing countries lack the infrastructure to support BEVs. So the BEVs would be scrapped early, shortening their working lives, thus reducing the lifetime over which their 'green' credentials are earned. (see Chap. 5 'Zero tailpipe emissions v lifetime emissions'), making them a less environmentally friendly option.

Chapter 9
Solutions

Abstract Chapter 9 examines solutions to overcome outstanding issues and considers developments in battery technology, range, range anxiety, weight, recharging, the cost of a BEV, solving the infrastructure problem and more practical implementation timescales.

Keywords Battery technology · Range · Range anxiety · Solid-state sulphide · BEV cost · Serial-drive PHEV · EREV · Implementation timescales

Developments in Battery Technology

Range

We saw in Chap. 4 'The issue of driving range' what has been achieved to date in respect of the driving range of BEVs and in 'Recent and ongoing battery developments' of the likely improvements that we may expect in the near future. Claimed BEV ranges of 600, even 1000+ km, and the promise of new technologies, such as solid-state batteries, promising much more, and safer, technology. That there are several different developments showing promise, lends to the probability that at least one, probably more than one, will succeed.

Even allowing for the fact that claimed ranges are rarely achievable in practice, that the vehicles with the longest range have the largest batteries so are therefore the most expensive—indeed, are usually the top-of-the-range, most expensive models in any case—and accepting that most average and small-sized BEVs on the market today have a reliable, useable range that is inadequate for anything other than usage as a second car or local runabout. In the 2030–2035 target period adopted by most governments, having researched the subject for this study, I think that there is good reason to believe, not certainty but good reason to believe, that the range issue is being addressed and that practical ranges will be achievable by all but the smallest BEVS.

B. Williams, *Electric Shocks: Are Drivers Rejecting the Electric Car?*,
https://doi.org/10.1007/978-3-031-98820-2_9

I stress that we are not there yet for most models, but we are close enough to remove range as one of the reasons that, in the longer term, will deter people from driving a BEV.

Range Anxiety

Range anxiety is a different problem to actual range limitation, and it is a far greater more intractable problem to BEV ownership and driving. It is a bigger, more intractable problem because it is psychological, with, in the case of BEVs, no cure. And to a large extent, it has little to do with the actual range achievable by any BEV.

When I was young, I got my first car because I was sent to work in a remote part of the country. There were three petrol stations within 50 miles, and they were only open a short number of hours in the day. I was always anxious about running out of petrol, so I relieved my range anxiety by always carrying a spare can of petrol in the boot. As far as I can remember, I never had to use it. But it was there just in case.

Later, when I decided to drive cross country from New York to San Francisco, intimidated by the vast distances involved in the USA, I packed a jerrycan full of gasolene in the trunk.

You can't do that with a BEV—a PHEV, yes. But not a BEV.

To the range-concerned, I would normally advise a long-range PHEV. To be 'greenest', a serial PHEV/EREV (see Chap. 5 'Plug-in Hybrid Electric Vehicles'), and recharge it regularly. That way the ICE engine will probably never cut in—but you can rest without range anxiety because it will be there if it is needed, and you won't find yourself stranded.

But under the current regulations they will be banned.

Weight

As we saw in Chap. 6 'The problem of weight', BEVs face several significant challenges related to their weight.

The core of the problem being that a BEV battery pack is heavy because of its technology.

Current BEVs rely on lithium-ion (or variants of) batteries in order to store the large amounts of electrical energy needed to propel around 2000 kg of motor car plus passengers and luggage for a practical driving range. The more energy a battery is required to store, the more cells it needs, which increases both its size and weight. But why are car batteries so heavy?

Lithium-ion batteries, the dominant battery type in BEVs, contain materials such as lithium, cobalt, nickel, and graphite, all of which are dense and heavy. The cathode materials (such as nickel-cobalt-manganese or nickel-cobalt-aluminium) are particularly heavy due to their chemical properties.

Lithium is a highly reactive metal that can self-ignite, especially when exposed to moisture (water or humidity in the air). When lithium comes into contact with water, it reacts to form lithium hydroxide and hydrogen gas, which can ignite spontaneously, particularly if the lithium is finely divided. Lithium can also react with the oxygen in the air, and when it becomes oxidised, it can heat up and potentially ignite if conditions allow.

Unlike metallic lithium, lithium-ion batteries do not contain pure lithium metal but use lithium compounds that are generally more stable. However, they can still catch fire or explode if damaged, overheated, or short-circuited. Lithium can self-ignite at relatively low temperatures (around 180–200 °C or 356–392 °F).

To ensure safety and longevity, BEV batteries are housed in reinforced casings to protect against impacts, overheating, and the fire hazards described above. These casings add significant weight to the battery pack.

In order to avoid the risk of overheating, BEV batteries require cooling systems (air-cooled or liquid-cooled) to manage temperature during operation. These systems include pumps, coolant, heat exchangers, and related components, which also add weight.

Manufacturers therefore have strike a balance between providing sufficient driving range and keeping the vehicle weight manageable. While adding more battery capacity increases range, it also increases weight, which can reduce efficiency and performance. This trade-off is a constant challenge in BEV design. A BEV battery pack is heavy in comparison to the rest of the weight of the vehicle, so the heavy vehicle requires more powerful motors to propel it. More powerful motors require more power which requires bigger batteries which add further to the weight, requiring even more powerful motors, and so on.

By the time the manufacturer has optimised, the battery pack is often accounting for up to 35% of the BEV's total weight.

Lithium-ion is one of the few battery technologies that can store enough energy source for a BEV, but it is far from ideal. Chap. 4 'Recent and ongoing battery developments' and Chap. 6 'The problem of weight' explore R&D efforts to find a better—lighter and safer—alternative. The most hopeful of which are solid-state batteries, which, through separate developments, Toyota and MG expect to productionise by 2028.

The Toyota initiative with Idemitsu focuses on sulphide solid electrolytes, which are seen as a promising material to achieve high capacity and output for BEVs. Sulphide solid electrolytes are characterised by softness and adhesiveness to other materials, which is suitable for battery mass production. Solid-state batteries using sulphide solid electrolytes offer significant advantages in terms of energy density, safety, and performance compared to conventional lithium-ion batteries. However, challenges related to cost, moisture sensitivity, and interface stability with lithium metal anodes, remain.

The MG commitment to solid-state batteries is also believed to be based on sulphide-based technology, as we know that the development is 'in-house' and that its parent company SAIC has been investing in research in this technology. But as of September 2024, that has yet to be confirmed.

Solid-state sulphide technology batteries promise to be lighter, more energy intensive and safer than lithium-ion, but will be more expensive and, at this point in time are still in the R&D laboratory.

The solution to the BEV weight problem is to change the battery technology—maybe that will start to happen by 2030, but it is too soon to be sure.

Until then, BEVs suffer from a weight problem which uses more energy, wears out components such as brakes and tyres more quickly (causing increased particulate pollution) damaging roads and bridges, and causing weight loading problems for parking facilities. .....in no small part ruining its 'green' credentials.

The problem of the weight of BEVs is very underestimated—its 'skeleton in the closet' (or perhaps more appropriately, 'the elephant in the room').

Recharging

For those who can recharge their BEVs battery with a charging point at home, recharging is a minor task whose only issue is remembering to do it, less disruptive (but more frequent) than driving to a service station to refuel an ICEV.

But there is no getting away from the fact that charging at a public charging point is an inconvenience and can be a major inconvenience. These issues were discussed in some detail in Chap. 8 'Availability of refuelling (home and away)' and 'Time to refuel'. In summary: First you have to locate a charging point that is available. Then you have to drive to it and hope that it is still available when you get there. Then you have to wait around for half an hour to an hour (longer in some cases) while your BEV battery recharges. Yes. Sometimes you can combine it with a meal break or shopping trip. But more often that is not the case. And while this may be an acceptable occasional event for those who normally recharge at home, is a real deterrent to BEV ownership for the 40% who cannot recharge at home.

As we have seen in Chap. 8, there is R&D into inductive charging while driving, but it is less energy-efficient (therefore more expensive for the user), would require a redesign of all BEVs, and, most significantly, massive investment in road infrastructure and payment systems from road authorities and governments who are currently struggling to install enough charging points to support 2% of the light vehicle park.

For 60% it may be an inconvenience they can live with. For 40% it is an intractable problem.

Cost of a BEV

As this work has alluded to in several places and detailed in Chap. 8 'Back-to-basics—What does "fit for purpose" for a family/personal vehicle require?—Purchase Price', a BEV costs a third more than an ICEV to buy outright. Maybe more than a third more to lease. A BEV costs more to insure than an ICEV. And this at a time when new car prices have risen about 25% over the last 6 years because of other factors such as regulatory changes, technological improvements, and inflation. [This means that the driver who last purchased an ICEV for £ 22,000 (US$ 27,500) 6 years ago would face a price of £ 27,500 (US$ 34,400) for a like-for-like ICEV replacement today, and a price of £ 36,600 (US$ 45,750) to replace it with an equivalent BEV—a price increase of some 66%.]

As we saw in Chap. 8, the price difference between a fuel tank for an ICEV which costs about US$ 50, and the battery for a BEV which costs US$ 12,000–US$ 15,000—maybe more for a long-range BEV, being the principal reason.

And, until a change in battery technology drastically reduces the price of a battery, the BEV, intrinsically, will be significantly more expensive than an ICEV.

Currently, EV battery costs are between US$ 140 kWh and US$ 160 kWh. The major OEMs are targeting US$ 100 kWh by 2015–2027 with economies of scale and new gigafactories coming online.

That will, if achieved, reduce the ICEV:BEV price differential to around 20%—a significant improvement, but still a marked price difference (some OEMs are claiming that $ 100 kWh would bring them closer to price parity than this).

(Note: Price disparity may not always be the case for a serial-drive PHEV, where a very small ICE engine recharging smaller batteries than a BEV would only require a modest improvement in battery cost per kWh to reach price parity with an ICEV.)

Solving the Infrastructure Problem

The issue of a hopelessly inadequate supply of publicly available EV charging points was addressed in some depth in Chap. 4 'The challenges of fuel supply (commercial infrastructure)'.

Using the EU as an example, in Chap. 4 we calculated that currently ICEVs are supported by some 92,000 'gas stations' each with an average of say eight filling points, i.e. say some 736,000 filling points. Assuming it takes 5 min per refuelling transaction (12 transactions per pump/hour). That gives a total capacity of servicing over 8.8 million refuels per hour. 8.8 million divided by the vehicle population of 280 million means that the refuelling architecture could potentially service (refuel) just over 3% of the vehicle population in any hour.

In order to provide the same level of support of 3% of 280 million electric vehicles, even at just 30 min per recharge would require 4.2 million ultra-fast charging points. Currently, the EU has just 96,000 ultra-fast chargers, indeed only 736,000 public charging points of all types.

And Chap. 4 concluded that *You can with justification argue about the figures all day long, but the fact that there is a hopelessly inadequate availability of public charging points in Europe to support a 100% BEV light vehicle population persists with any realistic numeric assessment.*

Another way to look at it is that in the EU there is one public charging point per every 360 vehicles, one ultra-fast charger per 3000 vehicles.

The issue in this section is not so much how do you calculate the shortfall, but what can be done about it?

The European Commission has set its 27 member states the lofty target 6.8 million public charging points by 2030. With 6 years to run, and a current total of 736,000 charging points, that is, a million extra charging points per year. But it has to rely on its Member States to make it happen. And how are they doing?

According to GridX 'Charging Report 2024', about 150,000 new public charging points were installed in the EU in 2023, but only 16% (24,000) of them were DC (ultra-rapid or rapid) charging points.

I leave it to the reader to assess whether the EU will reach its goal of 6.8 million by anywhere near 2030.

And I have to again pose the questions, 'why the hurry?' and 'why set unachievable targets?' if the electricity supplied won't be truly green before 2050.

More Practical Implementation Timescales

The reader will be aware that in Chap. 2 'The challenges of the fuel storage' I made the statement 'This work assumes the claims that electricity supplied from the network can and will become green'. I have tried to adhere to this belief throughout this work. But now we come to consider support infrastructure issues (and without getting into issues about what is green energy and what is not) the issue of 'when' electricity will become substantially 'green' sourced—not 100% green sourced, say (as a random indication) 80% green sourced—becomes important to the issue of BEVs contributing to decarbonisation. And here we are faced with facts and projections about objectives that we can't avoid and should be faced.

As of 2023, around 58% of the EU's electricity comes from hydrocarbons and 42% of the EU's electricity comes from renewable energy sources. The EU has set targets to increase this percentage, aiming for at least 45% green energy by 2030 (meaning 55% still from hydrocarbons). Projections suggest that by 2035 renewable energy could provide over 50% of its electricity and 60% of the EU's electricity by 2040. (source: Dialogue Earth).

Indeed, currently some 16% of EU's electricity is generated from coal, and even green conscious Germany will not phase coal power generation out before 2030.

To decarbonise the planet more quickly, it could be argued that electricity use should be *minimised* until at least coal fired generation is eliminated. The basis for that tenet is that governments/generators in all countries now give precedence to providing electricity firstly from renewable sources, then from natural gas, and in cases of peak demand have to fall back on their coal burning generation capacity (which is recognised as the worst offender in contributions to global warming).

It could be rationally argued that *until* coal burning is substantially phased out, say 2040 (the EU plans for coal to be almost entirely phased out by 2040 as part of its broader goal to cut emissions by 90% compared to 1990 levels), priority should be given to *avoiding electricity demand peaks* that force electricity generators to use their coal generation power stations.

Because by comparison:

- Coal-fired electricity power plants emit approximately 820 g of CO_2 per kilowatt-hour (gCO_2/kWh) of electricity produced.
- Natural gas electricity power plants emit approximately 450–500 gCO_2/kWh.
- The average family gasolene/petrol ICEV emits circa 260 gCO_2/kWh.
- (For a diesel car, circa 273 gCO_2/kWh.)

(source ChatGPT)

So (and I confess to this coming as somewhat of a shock to me) forcing drivers to use EVs, other than charging at off-peak times of low electricity demand (when the electricity is likely to be green-sourced) before the normal and peak rate electricity is largely 'green', far from cutting carbon emissions, loads the network and forces it to use non-green electricity generation which at best doubles CO_2 emissions per kWh, and at worse more than triples CO_2 emissions per kWh compared to driving a comparable ICEV.

Now this is ameliorated by the fact that a BEV electric motor converts its kWh more efficiently than an ICEV, but in most cases this is offset to a large extent by most BEVs having two electric motors to the ICEVs one ICE, and so the damage is not necessarily 2× or 3× worse. But it is worse!

And this then begs the question, *why if we are trying to reduce CO_2 emissions are we pursuing a policy that until about 2040 will increase CO_2 emissions caused by cars and far from combatting global warming, will contribute to its worsening? Surely a more selective policy would be more appropriate?*

If we were to pursue a more carbon-reduction friendly policy—such as encouraging those who can recharge off-peak to use BEVs, and encourage those who cannot (or have range anxiety) to use serial-powered PHEV/EREV or efficient modern hybrids, we would contribute far better to reducing CO_2 emissions and, coming back to the thread of support infrastructure, give countries a further decade to roll out adequate infrastructure—in the EU's case, 6.8 million charging points.

Chapter 10
Alternatives

Abstract Chapter 10 considers alternatives to the BEV. Social changes such as changes in driver demands/needs/behaviour, alternative fuels and alternative charging technology. The chapter considers rethinking the Hybrid and PHEV, addressing the on-street parking problem and makes a plea to stop viewing in black and white and imposing a one-size-fits-all solution.

Keywords Driverless taxi services · Traffic controls · FCEV · HICEV · Solid-state batteries · Megawatt chargers · Series PHEV system · On-street parking

Changes in Driver Demands/Needs/Behaviour

In Chap. 1, 'Social engineering, causes, and consequences' we saw the vision of the UN and the EC/EU to solve the contribution to CO_2 emissions from light vehicles by social engineering, i.e. changing peoples' behaviour. And in cities and other dense population areas much can indeed be done to improve public transport, pedestrian areas, cycleways, public transport information services, etc., and reduce reliance on, and use of, the car.

MaaS—Mobility-as-a-Service—where all the service providers cooperate in order to provide the optimum travel service/experience to the traveller—the dream of the social engineer—service provision attempts have failed because service providers do not want to expose their weaknesses to competitors, and a service provider would rather keep a traveller on its bus in a traffic jam than lose the fare to a competing mode of transport.

The dream of the SAEV—the shared automated electric vehicle—remains a dream as the practical use of automated vehicles proves more difficult than originally envisaged and prophesised by the auto manufacturers, and their dream of people switching from owning/leasing a vehicle to subscribing to an on-demand SAEV service is exposed for what it is basically—just subscribing to a taxi service, which they can do today and largely choose not to.

B. Williams, *Electric Shocks: Are Drivers Rejecting the Electric Car?*,
https://doi.org/10.1007/978-3-031-98820-2_10

Experiments with driverless taxi services have failed to impress, experiencing many challenges, including accidents, software glitches, and regulatory hurdles. They have struggled with handling unexpected situations such as road construction, emergency vehicles, or complex pedestrian interactions. Several reports of driverless taxis stopping unexpectedly because of a plastic bag blown onto the road or struggling to manage intersections have raised safety concerns and caused accidents, incidents, and resultant congestion delays.

In San Francisco, Cruise and Waymo vehicles have caused disruptions by stopping in the middle of the street due to technical issues, and misreading the traffic situation, especially where temporary measures were in place, confusing other drivers and causing traffic delays.

As a result, there has been significant pushback from local governments, particularly in major cities such as San Francisco, where concerns over safety have been raised.

(Tesla's 'Autopilot' and 'Full Self-Driving' (FSD) modes have also been involved in multiple incidents where the system failed to recognise obstacles or made inappropriate lane changes, leading to crashes.)

The same section in Chap. 1 also listed 12 good reasons why people benefit from having a car. Those who live in cities can usually commute to a fixed work location, or locations, by public transport, cycling, or walking, but many need a car to visit relatives and friends, do the school run and ferry children around their busy social and educational calendars. Others need their car in order to access their workplace, or carry out their work: carers, maintenance workers, social workers, doctors, professional advisors, some delivery drivers, private car hire and many taxi drivers, etc. And then there are holidays, excursions, and out-of-town shopping trips. So, cars remain an essential requirement for most citizens.

As pointed out in Chap. 1, many, indeed most, workplaces have also located themselves in the suburbs to avoid city centre congestion issues, and so one (or more) car(s) are essential for most families. The car represents the most convenient and comfortable means to achieve the journey, and for most journeys outside of the city centre, 'Mobility as a Service' multimodal journeys are a less attractive option.

The willingness to change travel behaviour is very questionable. Outside of city central areas, why would a car driving traveller who can easily, and largely comfortably drive his car from start point a), seamlessly to end point d) and back again, elect to choose to travel from a) to b), wait for a connection, possibly in inclement conditions, to catch a link to c) in order to wait for another connection possibly in inclement conditions, to catch a link to d), and then do the same thing to return later?

Many local authorities, failing to achieve their social engineering goals by benefitting the citizen are resorting to get their way by punishing the driver, by means of access charges, parking charges, onerous restrictions, and traffic controls, in order to achieve their goals.

I leave the reader to decide the merits of such strategies.

Yes. Social engineering can reduce CO_2 emissions in cities, but it will not eliminate them. And city centres are only a part of the problem, in any case.

Alternative Fuels

Chap. 5 'New Energy Vehicles' dealt with alternative fuels in its sections 'Fuel Cell Electric Vehicles (FCEV)' and 'Liquid Hydrogen ICE (HICEVs)'. The success of FCEVs will depend on two factors (a) a breakthrough in the energy cost of extracting hydrogen from water, and if so, (b) take-up of hydrogen as a fuel for heavy commercial vehicles.

(b) will roll out the required infrastructure and distribution systems and if it happens that would make FCEVs practicable for cars and light vans, offering an alternative to BEVs.

As we saw in Chap. 5, there are serious efforts and investment being made to achieve (a) by around 2030.

If (a) and (b) are achieved, FCEV becomes a viable competitor to BEV for zero CO_2 emission driving, and without many of its disadvantages.

Beyond 2030? Remember 'cyanobacteria' from Chap. 3?

Extracting hydrogen using cyanobacteria turns out (in the longer term) to be a promising approach to producing sustainable, clean hydrogen fuel by natural means, and if the challenges can be overcome, potentially at low cost.

Cyanobacteria use sunlight to split water (H_2O) into oxygen (O_2), protons (H^+), and electrons via photosynthesis. During this process, the protons and electrons can be combined to produce molecular hydrogen (H_2) under certain conditions. The hydrogen production is facilitated by enzymes such as nitrogenase and hydrogenase, which play key roles in reducing protons to hydrogen. Normally, these enzymes participate in nitrogen fixation or other metabolic pathways, but under hydrogen-producing conditions, they can be harnessed for hydrogen generation.

In nature, cyanobacteria often produce hydrogen only in low-oxygen or anaerobic environments since oxygen can inhibit the hydrogenase enzyme. Researchers are developing techniques to either create these conditions or modify cyanobacteria to be more efficient in hydrogen production even in the presence of oxygen.

(source: EU Carbon Brief)

Although cyanobacteria produce hydrogen, their natural efficiency is low. Genetic engineering and metabolic pathway optimisations are being explored to increase hydrogen output by reducing competition between photosynthesis and other cellular processes. While small-scale hydrogen production from cyanobacteria has been demonstrated, scaling the technology for industrial use is a big hurdle. Scientists are working on bioreactor designs that can support large-scale hydrogen production using these microorganisms. Some experiments have achieved promising increases in hydrogen yield through gene editing and optimised growing conditions.

(source ChatGPT)

So, in the longer term, the microorganisms that created the oxygen in the air we breathe may also be adapted to naturally produce the clean power that we need.

In respect of HICEV technology, as we summarised in Chap. 5 'Liquid hydrogen must be stored at extremely low temperatures (around −253 °C or − 423 °F) to remain in liquid form, which requires advanced insulation and cryogenic technologies……… more expensive to install and maintain, because of these insulation and cryogenic issues—making it difficult to deploy this technology at scale'. Because of these issues it is difficult to see how it could be rolled out to cars.

Butanol is not considered a viable alternative to BEVs because of the amount of agricultural land that it would use. Other biofuels are not considered an alternative fuel because they offer a means to decrease the CO_2 emissions of an ICEV, but do not remove such emissions.

Synthetic fuels (or e-fuels) are produced by combining hydrogen with carbon captured from the atmosphere, creating a carbon-neutral fuel. These fuels are chemically similar to gasoline or diesel, allowing them to be used in existing ICEVs. However, they are still in development, and their large-scale production is not yet cost-effective, so are not considered further in this work.

Alternative Charging Technology

One of the Achilles heels of the BEV is the time it takes to recharge the battery. And while this work concludes that new generation BEVs are finally solving the range problem, it comes via batteries of larger kWh capacity, which therefore take even longer to charge, thus exacerbating the time-to-charge problem.

What can be done about this problem?

In Chap. 4, 'Dynamic charging trials' showed detailed trials around the world (but particularly in Sweden) of technologies that enable vehicles to dynamically charge while on the move that seem to be achieving success although at lower conversion efficiency than conventional plug-in recharging. But the greatest obstacle to the practicability of these technologies is the infrastructure cost in their installation, maintenance, and operating costs and Chap. 4 concluded that this may be practicable for dedicated bus routes and similar limited operations, but though the technology can be made to work, the huge infrastructure costs involved render it an impractical solution for BEVs.

And so what other solutions to this problem can be found?

There is one light of hope that can be seen, and this work has already touched on it (see Chap. 4 'Recent and ongoing battery developments' and Chap. 6 'The problem of weight' and Chap. 9, 'Solutions') in the context of greater range, lighter weight, and safer battery technology: solid-state batteries.

The recharge time for solid-state BEV batteries is expected to be significantly faster than that of conventional lithium-ion batteries because of their higher energy density and better thermal management. However, the exact time to recharge will depend on the success in the commercial adaptation of this new technology's development and will also depend on the charger used to make the recharge.

When recharging away from home (where waiting-while-charging time is most inconvenient), according to DialogEarth and the EU's Carbon Brief, solid-state batteries are projected to support ultra-fast charging, potentially achieving an 80%

charge in 10–15 min. (This is much faster than typical lithium-ion batteries, which can take 30 min to 1 h for a similar charge on a high-powered DC rapid or ultra-rapid charger). BYD have recently demonstrated battery technology which they claim can achieve an 80% recharge in 5–10 min using 'megawatt' chargers and new battery technology. (no battery technology detail available at the time of going to print, but believed to be solid-state).

When recharging at home (where the waiting-while-charging delays are less commonly faced as recharging is generally undertaken on a planned overnight schedule) using slower AC home chargers, the charging time for solid-state batteries is likely to be similar to that for current BEVs but may still benefit from improved efficiency. Full charging could take 5–6 h, depending on the battery size and charger output.

Solid-state batteries recharge faster because they have a solid electrolyte, which typically exhibits lower resistance compared to liquid electrolytes, allowing for faster ion transfer, resulting in quicker charging. They also generate less heat during charging, which enables higher charging currents to be used without overheating or degrading the battery.

It must be remembered that solid-state batteries are still in the development and testing stages for mass production. But with companies like Toyota, Idemitsu Kosan, SAIC/MG, QuantumScape, Solid Power, Samsung, and potentially BYD actively developing solid-state battery technology, this is the most probable route to solve the time-to-recharge issue (and at the same time relieve the infrastructure shortage issue, as one rapid/ultra-rapid recharge point will then be able to service double or triple the number of customers).

A recharge time for a BEV of 10–15 min is only two to three times as long as the average tank refill time for an ICEV.

Recharging a series PHEV or series-parallel PHEV with a solid-state battery is expected to take 5–7 min because of the smaller battery capacity, giving ICEV refuelling a run for its money.

Stop Viewing in Black and White

To the reader who is looking for a single one-size-fits-all solution to outstanding BEV issues I say this 'You are in for a disappointment'.

There is no single solution because it is not a single issue.

Indeed, it is the naivety of the 'black & white' *we have a single problem—carbon emissions caused by cars and vans—simple solution 'make everyone drive BEVs because they have no tailpipe emissions'* approach that caused the mess that we are in, and getting deeper into by the day, and is causing disillusionment and rejection of EVs.

I say a 'mess' because to this authors opinion, it is this naïve and thoughtless single mindedness that if pursued according to plan, will ***increase demand for electricity faster than it can be supplied from 'green' energy*** sources, actually increasing the use of hydrocarbons, probably until 2040–2050. And worse, especially within the next decade, lead to peak electricity demands that fire up the use of the highly polluting coal-fired power stations, and causing deferment of their permanent closure. *Far from decarbonising the planet, this policy is adding to its CO_2 emissions and global warming!* A better, more sustainable, policy would surely be one that better matched electricity demand to that which can be supplied by environmentally friendly 'green' sources.

But there is also no single solution because there is also not a single entity called a 'driver'.

Drivers are a diverse lot, different drivers driving for different reasons, and the same driver driving for different reasons on different days, or parts of the same day; living in differing circumstances, with different budget means, priorities, and possibilities.

The optimum solution to minimise greenhouse gas emissions for the affluent family with a house and driveway, owning two cars, and who commutes on the train, will be different from the optimum solution to minimise greenhouse gas emissions for the care worker on minimum wage who lives in a flat, parks in the street, uses the car to go from one client to another.

The means to minimise GHG emissions vary from person to person and policy should inform and help them—even incentivise them—to make decisions that minimise GHGs. Not to force them into a unitary single policy just because it is good for some politician's soundbite.

Yes, the long-term solution is to use green energy, which for many may imply driving a BEV. For others, in the short/medium term they might make a bigger contribution to reducing carbon emissions by exchanging their gas-guzzling, emissions belching, but highly reliable 20-year-old Mercedes, for a second-hand but modern low emissions ICE hybrid or PHEV that they can just about afford, instead of looking at the brochures, deciding that they can't afford a BEV, and have nowhere to recharge it anyway, so give the trusty old Merc a pat and say 'you'll be good for another 5 years, won't you, my old friend'.

There is no single solution because as we have seen in the above chapters. BEVs don't have one problem—they have many unresolved issues.

Yes, governments do have to be proactive to reduce CO_2 emissions, but forcing the one solution on everybody is not the way to achieve that goal, and the use of BEVs in some situations, as we have seen in the above chapters, face issues that have to be resolved before they should ever be even recommended, let alone enforced.

There is no single solution because the real issue should not be about driving BEVs, it should focus on reducing emissions from car and vans, by whatever means; and to use the old saying 'There is more than one way to skin a cat'.

Rethinking the Hybrid and PHEV

The evolution of the ICE Hybrid and the Plug-in Hybrid Electric Vehicle (PHEV) is interesting to study.

Chap. 3 described the struggles, in the period 1960–2000, initiated largely by petroleum supply and cost issues of the 1970s, to develop viable electric vehicles. While the end results were not a great success in terms of a viable BEV, a lot of creative thinking and development took place to improve efficiency: notably regenerative braking which converts kinetic energy from braking into electricity, which is stored in a battery big enough to power electric motors; lithium-ion batteries; permanent magnet synchronous motors; and continually variable transmission.

The combined use of these features, and realisation of the efficiency of electric drive in slow-moving situations and for rapid acceleration, may not have led at the time to a viable BEV, but, combined with an ICE, led to the creation of the first commercial Hybrid car—the Toyota Prius—and a significant subsequent proliferation of hybrids by competing auto manufacturers.

This led, in 2007, to 'A123 Systems' and 'Hymotion' offering aftermarket plug-in conversion kits for the Toyota Prius, enabling it to operate as a PHEV and in 2009 led to the production model 'Toyota Prius Plug-in Hybrid'—A Prius with a larger battery and a plug-in charging system, enabling limited all-electric driving range, combining electric-only range with the flexibility of a gasoline engine for longer distances.

This approach became known as the 'Parallel PHEV System' where both the internal combustion engine and the electric motor can independently drive the vehicle. The system switches between or combines the two power sources based on driving conditions and battery charge levels.

General Motors, with the Chevrolet 'Volt' in 2011, and 'Spark' in 2013 started from a different perspective and took a different approach.

The Chevrolet Volt (and Spark and Malibu models) operated primarily as an extended-range electric vehicle (EREV). The Volt had a 16 kWh lithium-ion battery that was recharged by plugging into an external power source. When charged the electric motor powered the vehicle, and the gasoline engine remained inactive. However, when the battery became nearly depleted, the Volt seamlessly transitioned into extended-range mode, where a small 1.4-litre gasoline engine (acting as a generator) provided electricity to power the electric motor. In this mode, the gasoline engine did not directly drive the wheels but generated power to keep the electric motor running, allowing for a total driving range of about 300+ miles with a full tank of gas and a charged battery.

The Volkswagen i 3 uses a similar approach.

This approach became known as the Series Hybrid System, or Series PHEV System. More latterly becoming known as EREV.

But Chevrolet's conceptual thinking was limited to the idea of a small gasoline tank and small engine to top up the battery as a range extender in smaller EVs with limited space for the BEV battery. When longer range batteries became more common it switched its focus to the 'Bolt' BEV and discontinued the Volt/Spark/Malibu approach.

But elsewhere, particularly in China, with its more liberal 'New Energy Vehicles' approach they have been more adventurous in their conceptual thinking.

Typically, 'parallel' PHEVs have evolved from ICE hybrids—typically SUVs and estate cars—such as the Toyota RAV4/Lexus RX450, Mitsubishi Outlander and Prius—ICEV hybrids with large powerful engines, adding battery capacity and plug in recharge capability. The traditional large ICE provides the 'grunt' and the electric motors managing low-speed operations and providing a boost for rapid acceleration.

But BEV technology handles all aspects of driving. So what if you reverse the roles, and, as did the Chevy Volt, use the ICE engine simply to tick over at low steady revs to recharge the BEV battery? You don't need a 2, 3, or 4 litre ICE to perform that task. In most cases a,much smaller, more fuel efficient, low emissions ICE will do the job just fine, improving economy, and reducing emissions at the same time.

Auto manufacturers such as Li Auto have conceptualised this way (see Chap. 5 'China and PHEVs') and recently launched a full-sized SUV PHEV, the L9 (described by the company as the pinnacle of large luxury SUVs, with what it says is a spacious interior specifically for Chinese three-generation family households). The automaker says the model offers passengers more room than other luxury automaker offerings with an ICE engine of just 1.5 litres by using this EREV approach, which is now becoming known as the 'Extended Range Electric Vehicle'.

The Li 9 has two electric motors: a 130-kilowatt (174 hp) motor in the front and a 200-kilowatt (268 hp) motor in the rear. It is also equipped with a front-mounted 1.5-litre turbocharged four-cylinder petrol engine with a 65-litre (17.2 US gal; 14.3 imp gal) petrol tank capacity. The petrol engine is a range extender for the electric motors; it does not directly power the wheels.

Li Auto claims a range of 1315 km (817 miles), and an electric-only range of 215 km (134 miles). The L9 accelerates 0–100 km/h (0–62 mph) in 5.3 s.

The vehicle (also packed with ADAS and infotainment) is on the market in China at RMB 459,800 (circa US$ 68,500/£ 54,000/€ 63,000), a price that the automaker claims is lower than any other similar SUV on the market. For comparison, in China the BMW X7 and the Mercedes-Benz GLS crossovers start at RMB 1 million (US$ 149,000) and RMB 1.07 million (US$ 159,000), respectively.

(source Li Auto, Wikipedia, and Technode.com)

Its smaller Li 8 model is designed using the same principles. (see Chap. 5 'China and PHEVS').

In 2023, Li Auto sold a total of 376,030 vehicles (Li 7, Li 8, Li 9 models) (source CarNewsChina.com), so this approach is well proven and marketable.

This work suggests that in order to reduce CO_2 emissions most rapidly, other manufacturers, particularly PHEV manufacturers and conventional hybrid manufacturers, should adopt this EREV approach ASAP. Indeed, BEV manufacturers should be considering it as an option to add to their range.

So far, at the time of writing, Nissan has taken up this approach with its 'ePower' variants of some of its models (Qashqai and X-Trail), but have not yet provided a battery of significant range to be a true EREV).

Time for other OEMs to open their eyes to this course ……. But of course regulations forcing the use of BEVs unhelpfully cut against this approach in some countries.

Addressing the On-Street Parking Problem

Readers who have fought their way through this work this far will by now be clear on its position regarding those who park on the street:

(a) 40% of homes do not have a garage or driveway. Meaning that probably in excess of 25% + of cars and light vans are parked on the street.
(b) Due to the random nature of most street parking, do not have the opportunity to recharge at home.
(c) The majority of the drivers of those vehicles are in the lowest quartile in respect of income.
(d) Even if there was an adequate number of recharging points, which there is not, nor ever likely to be, in general on-street parkers could not afford the premium rates charged for rapid/ultra-rapid DC charging.

For much of these 40% of homes a BEV is simply not a viable option.

As was said in 'Stop viewing in black & white' above '…….. Yes, governments do have to be proactive to reduce CO_2 emissions, but forcing the one solution on everybody is not the way to achieve that goal, and the use of BEVs in some situations, as we have seen in the above chapters, face issues that have to be resolved before they should ever be even recommended, let alone enforced'.

And from that subchapter too, the observation/recommendation '…. For others, in the short/medium term they might make a bigger contribution to reducing carbon emissions by exchanging their gas-guzzling, emissions belching, but highly reliable 20 year old Mercedes, for a second-hand but modern low emissions ICE hybrid or PHEV that they can just about afford, instead of looking at the brochures, deciding that they can't afford a BEV, and have nowhere to recharge it anyway, so give the trusty old Merc a pat and say 'you'll be good for another 5 years, won't you, my old friend'.

Chapter 11
Conclusions

Abstract Chapter 11 summarises the conclusions the author draws as a result of the study, and the author points out that the previous chapters have done their best to objectively present facts and events. Chap. 11 represents the authors' conclusions—the reader may well draw different conclusions.

Keywords ZEV regulations · New energy vehicles · Series-PHEVs · Cost-burden · End-of-life recycling · WEEE · Cartridge-based HICEV · EREV

Throughout the chapters of this work I have tried to remain objective at all times. It may have seemed to the reader from time to time that I was negatively trying to find fault with BEVs. This work has tried to identify solutions to BEV problems, and that by definition means firstly identifying the problems. I have to the best of my ability used the objective approach that has served me well in over 40 years of consulting:

The four 'W's questions:

- Where are we?
- Why and how did we get here?
- Where are things heading?
- What is the real issue?

And the four 'F's resolutions:

- Find the problem(s).
- Face the problem(s).
- Find the solution(s) or alternatives.
- Futureproof the results.

And this work has identified that there are problems with BEVs—multiple issues that require change.

Are drivers rejecting the electric car? This work has shown that some are not, but some are (and mainly for good reasons).

B. Williams, *Electric Shocks: Are Drivers Rejecting the Electric Car?*,
https://doi.org/10.1007/978-3-031-98820-2_11

But whereas this work to this point has made best efforts to objectively establish the facts—the four Ws and the four Fs for as many aspects of BEVs as possible, from which the reader has enough information to objectively make their own judgments and draw their own conclusions, this chapter—Conclusions—is different, in that it presents the author's opinion based on the studies that I have undertaken.

So whereas this work to date has identified issues, found solutions, alternatives, and apparently unsolvable problems, and passed some observations, this chapter provides the conclusions and recommendations from the author's viewpoint, gathered as a result of undertaking this study. You may or may not agree with all of them. Some of them were a surprise to the author.

The first conclusion and the biggest surprise is this: the move to drive BEVs is part of the drive to reduce human-generated carbon emissions (that it is believed are a prime cause of global warming). It is estimated that car and light vans produce 14% of GHG emissions. (meaning that 86% are generated by other sources, the most significant of which being the generation of power 42%. But target dates to eliminate that emissions source [power generation] are for most of the world 2050, and for the biggest polluter, China, 2060. Currently only around 30% of generating capacity in the USA is generated from renewable sources; Europe, around 42%; China around 35%; and in the UK 51%. The forecasts are that it will be 2040 or later before even 60% will be generated from renewables).

Until most, say 80%+, energy is green sourced, and it could be rationally argued that until coal burning is substantially phased out—say around 2040–2045 (the EU plans for coal to be almost entirely phased out by 2040 as part of its broader goal to cut emissions by 90% compared to 1990 levels), priority should be given to *avoiding electricity demand peaks* that force electricity generators to use their coal generation power stations. That is to say, to minimise the peak loads on the biggest polluter—which is electricity power generation.

BEVs consume electricity. Encouraging more BEVs increases demand for electricity.

I deliberately restate here the critically important ecological point that I made in the last chapter. Requiring BEV use by 2035, even worse 2030, will increase demand for electricity faster than it can be supplied from 'green' energy sources, actually *increasing* the use of hydrocarbons, probably until 2040–2050. And worse, especially within the next decade, lead to peak electricity demands that fire up the use of the highly polluting coal-fired power stations, and causing deferment of their permanent closure. *Far from decarbonising the planet, this BEV only policy is adding to its* CO_2 *emissions and global warming!* A better, more sustainable, policy would surely be one that better matched electricity demand to that which can be supplied by environmentally friendly 'green' sources. The current policy defeats the very logic of switching over to BEVs (to improve the environment and slow global warming), and directly harms the environment.

The consequences (of 100% switchover to BEVs before 2040 [5 years after 2035 cut-off date for ICEV sales]) would likely be power cuts, electricity rationing, and increases in electricity prices.

The unavoidable conclusion is that EVs should be encouraged only where they can be recharged off-peak, and the counterproductive ZEV regulations and penalties should be scrapped entirely, or at least deferred until after 2040.

The second conclusion is that the Chinese 'New Energy Vehicles' policy is more likely to persuade drivers to move to a vehicle that causes lower GHG emissions more quickly.

I will go further and suggest that *as the objective is to cut GHG emissions, drivers should be incentivised to do so.* I do not propose restoring grants to purchase BEVs (for the logic above, that could be counterproductive), but a simple measure like waiving a year's car tax if the replacement vehicle has 25% less tailpipe emissions than the vehicle it replaces. Or a simple, easy to claim, cash grant of the same scale if it does.

Measures like these would benefit not only BEV purchasers, but would benefit those downsizing ICE engines, moving from ICE to hybrid or PHEV, and is likely to be a good way of reducing the number of polluting old bangers being driven. To this authors' mind, such measures would be a more efficient way to reduce GHG emissions in the next 15–20 years (until electricity generation is largely generated from green sources).

A conclusion that has rung out clearly many times in the above chapters is that governments have to come to terms with and accept that for a substantial proportion of the population—those who park their vehicles on the street and do not have the opportunity to recharge at economy tariff rates overnight (most of 40% of homes in the case of the UK)—a BEV is often not a practical option. This has long-term implications. This sector is largely serviced by second-hand vehicles, so in the longer term needs a continuing supply of low emission non-BEV second-hand vehicles, such as small engine hybrids and PHEVs, preferably small IC engined series-PHEVs an EREV, (described above)—and that implies that *part of the market for new vehicles (probably largely the rental sector [who are currently struggling to sell-on used BEVs]) needs to be EREVs, PHEVs and small engine hybrids, not BEVs, in order to minimise GHG emissions most rapidly.*

A problem that has received little attention yet (because so far less than 3 cars in every 100 are BEVs) is the impact (I choose my words carefully) that BEVs have on the *roads and parking infrastructures*, especially in Europe and the far east.

Outside of North America, the trend to larger cars and SUVs is already impacting roads and causing pot-hole damage. BEVs weigh in a third heavier, largely because of the weight of the battery packs, but also the vehicles tend to be larger in order to provide space for the battery packs.

In time, as the proportion of BEVs increases, this will cause stress problems for multistorey car parks, designed for smaller, lighter, vehicles; and exacerbate the pot-hole problem.

The only solutions are for the auto manufacturers to find ways to reduce the weight of BEVs (see next conclusion) or drivers to opt for smaller vehicles. I am not in favour of restrictions, but local authorities will be forced to introduce size/weight restrictions for safety reasons, reduce the number of parking places by increasing bay widths (reducing capacity), or close parts of their car parks.

An important conclusion, with great implication for the auto manufacturers, is the consequences of the *cost burden* on the driver of the cost of BEVs, and the consequences that fall on the auto-manufacturing sector as a result.

The relationship between car prices and average family income has evolved over time, varying by country and decade:

In 1990, the average price of a new 'family' car in Germany was about €15,000–€20,000, while the average family income was around €22,000–€26,000, leading to a car-to-income ratio of roughly 70%–80%.

In the USA (in 1990), new 'family' car prices averaged around $16,000–$17,000, while the median household income was approximately $30,000, meaning a car cost about 55%–60% of annual income. And in the UK (in 1990}, the average new 'family' car cost around £9000–£10,000, with a median household income of about £16,000 - £18,000, giving a car-to-income ratio of roughly 55%–60%.

By 2000, in Germany, 'family' cars cost around €18,000–€25,000, and household average incomes were about €30,000–€35,000, improving the ratio to about 60%–70%, increasing the incentive to renew or upgrade vehicles, and meaning good times for the auto manufacturing sector.

In the USA, by 2000, 'family' car prices had risen to $20,000–$22,000, with household income around $42,000, improving the ratio to around 50%. In the UK, by 2000, 'family' car prices had increased to around £12,000–£14,000, with household incomes around £22,000–£25,000, bringing the ratio to around 50%–55%—again good news for the auto manufacturers.

Now step forward nearly a quarter of a century to 2024. 'family' cars have got a lot more sophisticated and safer–but at a cost, and the financial impacts of the banking crisis of 2008, the effects of Covid and the war in Ukraine have stretched family budgets worldwide.

By 2024, German 'family' car prices rose to around €40,000–€45,000, with the average household income roughly €55,000, making the ratio close to 70%–80%, taking German families back to where they were in 1990—resulting in delayed replacement of vehicles, and moving downmarket: bad news for the auto manufacturers.

By 2024, the average US 'family' car price has jumped to $45,000–$49,000, while median income reached roughly $75,000, making a car about 64% of household income, a decline of nearly 15%. In the UK 2024, average 'family' car prices stand at around £30,000–£33,000, while the median household income was £47,000–£50,000, bringing the ratio close to 65%–70%. Again a decline of 15%, with the same result of delayed replacement of vehicles, and moving downmarket: bad news for the auto manufacturers.

All of which leads to crises and rationalisations in the automotive industry.

And note, I have not mentioned the word BEV yet in this section (although 20% of new car sales in 2023/4 are BEVs, and form part of the reason for the price hike).

Now come back to the family car purchase of (and I deliberately use the lower end figures of) €40,000/$45,000/£30,000—representing 72%/60%/63% of family annual income. Or the family make the choice to purchase the BEV equivalent at 30% + more, say €50,000/$57,000/£39,000–90%/76%/83% of family income—and the reason for reticence to buy/lease a BEV is clear—with the same result of delayed replacement of vehicles, and moving downmarket, or, replacing with a less expensive ICE vehicle: bad news for the auto manufacturers.

All of which leads to more crises and rationalisations in the automotive industry who, in Europe and the UK, are required to manufacture a certain % of BEVs, whether the demand is there or not, or face stiff penalties.

This seems like a good policy to destroy an industry that has been part of the bedrock of affluence of western countries post-WWII. And furthermore, delays the use of less polluting vehicles.

Another conclusion drawn by the author as a result of this work is that although *lithium* was a good transitional technology to produce a first viable BEV battery, it is a bad long-term choice. The metal is expensive and politically insecure to process, is unstable, exothermic, and liable to 'thermal runaway'—as we have seen in several BEV fires following crashes, and many, many examples of electric bike and scooter fires. At the end of the battery life it is hard, dangerous, and uneconomic to dispose of/recycle. The resultant measures to keep it stable when operating add to system complexity, weight, and operational inefficiency.

And we recently find out from Israeli actions in Lebanon with remotely exploded pagers and walkie-talkies, as cars are now 'connected' devices, it is a security risk if a mal-intentioned bad actor remotely were to deliberately instigate a thermal runaway event (in one or many vehicles). Lithium batteries are not the only new risk in that department but deliberately introducing an exothermic thermal runaway risk into every car does not seem to be a wise strategy for national security in an increasingly unstable world.

Cobalt is often used in lithium batteries. It too is a bad long-term choice. Cobalt is expensive and politically insecure to process, is toxic, and is a recognised carcinogen.

I conclude that so long as the current lithium technologies are used, insurers will have to play safe and write off battery packs involved in a collision. This may involve writing off the vehicle as a result. *Insurance premiums* for BEVs will remain higher than insurance premiums for ICEVs. When battery technology changes to a more stable material (e.g. solid-state batteries), the situation should even itself out. But the higher price of most BEVs will mean that some differential will remain.

Something more has to be done about *end of life recycling*, which is technically possible, but uneconomic. Not many BEVs, except those that have crashed, have yet reached that point. Regulations in Europe (the WEEE [Waste Electrical and Electronic Equipment] Directive) pass that responsibility to the OEM, but this is avoided in many cases by the vehicle being shipped to a less regulated third-world country before it reaches end of life, and well-established profitable vehicle recycling businesses in the home country.

Both the USA and China have regulatory frameworks that are similar to the WEEE Directive. China has a national framework called the 'Regulation for the Administration of the Recovery and Disposal of Waste Electrical and Electronic Products', also known as the 'China WEEE'. This regulation requires producers to handle e-waste collection and recycling. It establishes a fund system, where manufacturers and importers contribute to the e-waste management fund, used to subsidise certified recyclers.

The USA does not have a federal equivalent to the WEEE Directive, but many states have enacted e-waste recycling laws. The main framework is based on 'Extended Producer Responsibility' (EPR), where manufacturers are responsible for recycling electronics (e.g. California *'Electronic Waste Recycling Act'*; New York State *'Electronic Equipment Recycling and Reuse Act'*).

But on the ground, there are, as yet, few established BEV battery recycling facilities. That has to change.

A thread running through this work has been the *driving range of a BEV*. I conclude as a result of this work that in respect of current BEVs on the road, the driving range of most current BEV offerings is now adequate for most users, but not adequate for others. However, generally speaking new models coming on-stream or already available have addressed this problem, but at the cost of longer times required for recharge and increased further the cost of BEVs.

While the actual problem of range may have been addressed, I conclude that, for the reasons expounded in Chap. 9, the issue of *range anxiety* is intractable and is best addressed by driving a PHEV, preferably a serial-PHEV (EREV). Better still, drive an EREV. That does not fit well with current ZEV regulations in some countries, but an extended range electric vehicle – EREV – even for an SUV or sports car EREV, cuts off the demand for a large gas-guzzling IC engine (an EREV gets its acceleration from its electric motor(s)), cutting emissions at a stroke, enabling zero tailpipe emissions for most city driving, removes range anxiety, and enables full BEV driving where green electricity is available. That, outside of China, EREVs are barely promoted, or even available, is an anathema to this author, and in contention with the most practical means of achievement of declared targets to minimise emissions.

The *availability of recharge points* is barely keeping pace with demand; if the 2035 targets for BEV take up are achieved, it is likely to be hopelessly inadequate. A change to more rapid charging solid-state batteries may alleviate the situation.

Much of the argument given to drivers to justify the additional price they have to pay in cash or lease payments for a BEV is that the refuelling costs are lower. But I conclude that logic to be short term and flawed because it is based on the premise that gasolene/diesel carries a 50%–120% tax burden, while electricity is available at no tax or very low tax. The longer term reality is that the duties/taxation on gasolene/diesel form a significant part of government income, and if it is not collected because vehicles have gone electric, governments will have to generate that income from elsewhere (such as pay-per-kilometre tax or monthly road use tax). i.e. the BEV driver will not avoid that tax burden much longer.

If the ambitions to achieve low-cost green hydrogen succeed, and it is beginning to look hopeful that they will, then two additional viable alternatives to BEVs will emerge:

(a) FCEV (fuel cell electric vehicles).
(b) Cartridge-based HICEV (hydrogen internal combustion engine vehicles).

It may be that FCEVs can also be powered from hydrogen cartridges.

Hydrogen cartridges seem to solve the problem of infrastructure roll-out for hydrogen, eliminating the need for forecourts to install hydrogen pumping points. The light weight of hydrogen stands to its favour.

Details of the financial viability of the solution have yet to emerge, and further study of safety aspects studied. But if the supplies of low-cost green hydrogen materialise, this cartridge solution could offer a viable alternative to BEVs in the future. If successful, Governments need to ensure that their regulations allow for it, as it will solve many of the issues associated with BEVs (though BEVs will still form part of the solution).

Recommendations in a Single Sentence.

Scrap ZEV and 2035 targets; and at least until 2040 encourage/incentivise EREVs (Serial PHEVs) and only encourage BEV use where it can be recharged at period of low electricity demand (e.g. overnight); encourage/incentivise solid-state batteries and ban lithium batteries as quickly as practicable; explore the practicality of cartridge based hydrogen, and for national security reasons ensure national regulations control vehicle communications (not 4/5/6 g industry practices alone).

Bibliography[1]

1979 World Climate Conference https://unfccc.int/cop3/fccc/climate/fact17.htm
A123 Systems https://www.a123systems.com/
A35 Brebemi https://www.brebemi.it/
AAA Consumer Survey https://newsroom.aaa.com/2023/11/annual-electric-vehicle-sentiment-survey/
BB https://global.abb/group/en
GAC's Aion LX plus https://electrek.co/2022/01/05/gac-aion-begins-sales-of-lx-plus-suv-with-1008-km-626-miles-nedc-range-for-just-over-72000/
ACC (a joint venture between TotalEnergies and Stellantis) https://totalenergies.com/infographics/acc-totalenergies-stellantis-and-mercedes-benz-joint-venture
Acura ZDX https://www.acura.com/suvs/zdx
ACWA Power https://acwapower.com/
Aion Hyper SSR https://aionhk.com/en/aion-hyper/

[1] **About this bibliography**

Every organisation, named person, vehicle model, product name, Convention, Regulation, used in this work is explicitly alphabetically referenced in this bibliography and provided with an internet link to where more information/ source of information can be found. The author hopes you find it comprehensive.

By 'explicitly alphabetically referenced' is meant 'as it appears in the text'. (so if the body text states 'Ányos Jedlik' the biblio reference will appear at Án [and not 'J']). Vehicle models may be referenced under the model name or manufacturer/model, depending on how it first appears in the text.

In many cases, in order to have some consistency in presentation in the text, descriptions (especially re vehicle models, production data etc.) have been compiled from multiple sources, so specific biblio sourcing referencing is not appropriate. But wherever possible the reader is referenced to a source where the manufacturers/source data can be obtained, albeit presented in a different format.

See the Caveats at the start of this work.

If the biblio reference is to a link other than a manufacturer/obvious source this is because the author has not been able to acquire the required data from originator. This may be because the source no longer exists, predates the internet, or no longer available/couldn't be located on the originators site.

B. Williams, *Electric Shocks: Are Drivers Rejecting the Electric Car?*,
https://doi.org/10.1007/978-3-031-98820-2

Aion LX https://cnevpost.com/2021/11/19/gac-aion-officially-launches-aion-lx-plus-with-range-of-1008-km/
Aion S https://insideevs.com/news/343627/gac-aion-s-is-equipped-with-588-kwh-catl-ncm-811-battery/
Aion V https://www.gac-motor.com/en/media/newsdetail/id/352.html
Air Products https://www.airproducts.co.uk/
Aiways U5 https://www.ai-ways.eu/en/u5/
Aiways U6 https://www.ai-ways.eu/en/u6/
AMC 'Electron' https://www.electricvehiclesnews.com/Footer/History/Companies/AMCelectron.html
AMC 'Amitron' https://en.wikipedia.org/wiki/AMC_Amitron
American Motors Corporation https://automotiveamerican.com/2024/09/15/the-rise-and-fall-of-american-motors-corporation-amc-1954-1980s/
Amprius https://amprius.com/
Ányos Jedlik https://en.wikipedia.org/wiki/%C3%81nyos_Jedlik
Arcfox α-S HI https://www.baicgroup.com.cn/en/news/22.978.html
https://www.huaweicentral.com/huaweis-technology-arcfox-alpha-s-series-highlights-specification-price-availability/
Asian Renewable Energy Hub https://en.wikipedia.org/wiki/Australian_Renewable_Energy_Hub
Audi https://www.audi.com/en.html
Audi 'Duo III' https://www.audi-technology-portal.de/en/mobility-for-the-future/hybrid-vehicles/audi-duo_en
Audi e-tron https://www.audi.co.uk/uk/web/en/electric/e-tron.html
Audi Q4 e-tron https://www.audi.co.uk/uk/web/en/electric/e-tron.html
Audi Q4 Sportback e-tron https://www.audi.co.uk/uk/web/en/electric/e-tron.html
Audi Q5 e-tron https://www.audi.co.uk/uk/web/en/electric/e-tron.html
Aulton https://cnevpost.com/2022/08/12/aulton-says-it-has-achieved-30-million-cumulative-battery-swap-services/
Auto Express https://www.autoexpress.co.uk/
Automated Vehicles and MaaS: Removing the barriers' Wiley/IEEE ISBN: 9781119765349 https://ieeexplore.ieee.org/book/9466288
Avis Budget Group https://www.avisbudgetgroup.com/
BAIC https://www.baicglobal.com/
BAIC battery swap https://bcf-auto.com/pages/battery-swap
BluePark https://en.wikipedia.org/wiki/BAIC_BluePark
BAIC EU300 https://torellos.en.made-in-china.com/product/MEwrXUQTOaRd/China-2024-New-Electric-Car-Baic-EU300-Electric-Vehicle-EV-Car.html
BAIC EU5 https://www.baicgroup.com.cn/en/passenger-vehicle.html
Beardsley Electric Company https://en.wikipedia.org/wiki/History_of_the_electric_vehicle
Beijing Electric Vehicle Co. Ltd. (BJEV) https://en.wikipedia.org/wiki/BAIC_BluePark
Beijing EU5 https://www.chinapev.com/specs/2020-baic-beijing-eu5-ev-technical-specs/#google_vignette
Beijing EU7 https://www.chinapev.com/specs/2019-baic-beijing-eu7-ev-technical-specs/#google_vignette
Beijing EX5 https://carnewschina.com/2018/02/28/beijing-auto-ex5-compact-electric-suv-400-kilometer-range/#google_vignette
Beijing Institute of Technology https://english.bit.edu.cn/
Benben 'E-Star' https://www.chnecar.com/news/chinese-electric-car-news/from-78-900-yuan-the-2023-benben-e-star-colorful-models-are-listed-with-a-range-of-310km/
Benjamin Franklin https://en.wikipedia.org/wiki/Benjamin_Franklin
Bentley https://www.bentleymotors.com/en.html
Bestune E01 https://db.carnewschina.com/bestune/bestune-e01/2020
Bestune E05 https://bestune.jo/

Besturn B30 EV https://carnewschina.com/2017/10/24/new-besturn-b30-ev/#google_vignette
BMW https://www.bmw.com/en/index.html
BMW 1602 Elektro-Antrieb https://en.wikipedia.org/wiki/BMW_1602_Elektro-Antrieb
BMW AG https://www.bmwgroup.com/en.html
BMW E1 https://en.wikipedia.org/wiki/BMW_E1
BMW Group https://www.bmw.com/en/index.html
BMW Hydrogen 7 https://www.press.bmwgroup.com/united-kingdom/article/detail/T0014470EN_GB/the-bmw-hydrogen-7
BMW i3 https://www.bmw.co.uk/en/electric-cars.html
BMW i4 https://www.bmw.co.uk/en/electric-cars.html
BMW i5 https://www.bmw.co.uk/en/electric-cars.html
BMW i7 https://www.bmw.co.uk/en/electric-cars.html
BMW Ix https://www.bmw.co.uk/en/electric-cars.html
BMW iX xDrive50 https://www.bmw.co.uk/en/electric-cars.html
BMW iX1 https://www.bmw.co.uk/en/electric-cars.html
BMW iX3 https://www.bmw.co.uk/en/electric-cars.html
BMW X7 https://www.bmw.co.uk/en/electric-cars.html
BP https://www.bp.com/
BP Pulse https://www.bppulse.co.uk/
British Vehicle Rental and Leasing Association (BVRLA) https://www.bvrla.co.uk/
Brittanica https://www.britannica.com/
BYD (a battery manufacturer) https://www.bydglobal.com/en/NewEnergy.html#:~:text=BYD%20is%20the%20world%27s%20leading%20producer%20of%20rechargeable,consumer%20electronics%2C%20new%20energy%20vehicles%20and%20energy%20storage.
BYD 'Dynasty' model range https://www.bydauto.com.cn/pc/
BYD 'Ocean' Series https://carnewschina.com/2021/09/08/apart-from-dolphin-byds-ocean-series-will-include-seal-sealion-and-seagull/#google_vignette https://www.byd.com/eu/car/ocean
BYD 'Qin' https://www.byd.com/eu/car/qin https://carnewschina.com/2024/05/28/byd-qin-l-launches-in-china-with-fifth-generation-phev-system/
BYD 'Tang' https://www.byd.com/eu/car/tang
BYD Atto 3 / Yuan Plus https://www.byd.com/eu/car/atto3
BYD Auto https://www.byd.com/en/about-byd
BYD Auto Industry Co. Ltd. https://www.byd.com/en/about-byd
BYD Dolphin https://www.byd.com/eu/car/dolphin
BYD e6 https://sg.byd.com/wp-content/uploads/2019/11/e6_brochure_2018.pdf
BYD F3e https://en.wikipedia.org/wiki/BYD_F3
BYD Han https://www.byd.com/eu/car/han
BYD Seagull https://www.byd.com/eu/car/seagull
BYD Song Plus DM-I https://www.byd.com/en-sa/car/song-plus-dmi-en
BYD Tang DM https://www.byd.com/eu/car/tang
BYD Yuan Plus https://www.byd.com/content/dam/byd-site/america-public/flyer/yuan-plus-flyer-en-20230525.pdf
Cadillac https://www.cadillac.com/
Cadillac Celestiq https://www.cadillac.com/electric/celestiq
Cadillac Lyriq https://www.cadillac.co.uk/en/electric/lyriq
CALB http://en.calb-tech.com/
California 'Electronic Waste Recycling Act' https://calrecycle.ca.gov/electronics/statutes/#:~:text=Learn%20about%20the%20law%20that%20established%20a%20funding%20system%20for
Camille Jenatzy https://en.wikipedia.org/wiki/Camille_Jenatzy
CARB California Air Resources Board https://ww2.arb.ca.gov/
CarNewsChina.com www.carnewschina.com

CATL https://www.catl.com/en/
Chalmers University in Sweden https://www.chalmers.se/en/
Changan Automobile https://www.globalchangan.com/
Changan Avatr 11 https://www.globalchangan.com/avatr.html
Changan BenBen EV https://db.carnewschina.com/changan-oshan/changan-ben-ben-e-star/2023
Changan CS15 E-Pro https://www.changan.com.cn/cars_m/E-pro/index.html
Changan CS15EV https://www.changan.com.cn/cars_m/E-pro/index.html
Changan CS55 E-Rock https://www.changan.com.cn/cars_m/E-pro/index.html
Changan Lumin https://www.changan.com.cn/cars_m/E-pro/index.html
Changan UNI-K Idd https://www.changan.com.cn/cars_m/E-pro/index.html
Changan's "6321" new energy technology goals, https://www.globalchangan.com/newsroom/infinite-evolution-everlasting-success-changan-automobile-2023-sales-surpass.html
Charles David Keeling https://en.wikipedia.org/wiki/Charles_David_Keeling
Charles Kettering https://www.britannica.com/biography/Charles-F-Kettering
Chat GPT https://openai.com/index/chatgpt/
Chery Arrizo 5e https://www.chery-eg.com/en/car/arrizo-5
Chery Automobile Co. Ltd. https://www.cheryinternational.com/
Chery Fulwin T10 https://carnewschina.com/2024/07/26/chery-fulwin-t10-entered-market-price-starts-at-26300-usd/#google_vignette
Chery Omodo E5 https://cherymotor.com.au/models/omoda-e5
Chery QQ3 EV https://db.carnewschina.com/chery-new-energy/chery-qq3-ev/2012/params
Chery Tiggo 8 Plus PHEV https://www.cheryinternational.com/pc/models/tiggo/tiggo8proe/index.shtml
Chevrolet https://www.chevrolet.com/
Chevrolet Blazer EV https://www.chevrolet.com/electric/blazer-ev
Chevrolet Bolt EUV https://www.chevrolet.com/electric/lineup
Chevrolet Bolt EV https://www.chevrolet.com/electric/lineup
Chevrolet Equinox EV https://www.chevrolet.com/electric/lineup
Chevrolet Malibu https://en.wikipedia.org/wiki/Chevrolet_Malibu
Chevrolet S-10' EV https://www.chevrolet.com/electric/lineup
Chevrolet Spark EV https://www.evspecifications.com/en/model/89aeb1
Chevrolet Volt https://en.wikipedia.org/wiki/Chevrolet_Volt
China Association of Auto Manufacturers http://en.caam.org.cn/
China Automotive Power Battery Industry Innovation Alliance http://en.caam.org.cn/Index/show/catid/43/id/1736.html
China WEEE - Regulation for the Administration of the Recovery and Disposal of Waste Electrical and Electronic Products https://english.mee.gov.cn/Resources/laws/regulations/Solid_Waste/201712/P020171213581119365936.pdf#:~:text=policies%20a
Chinese 'Auto Industry Restructuring and Revitalization Plan' https://www.cbinsights.com/research/china-electric-vehicles-expert-intelligence/#:~:text=In%202009%2C%20the%20Chinese%20central%20government%20approved%20the,option%20for%20rapidly%20urbanizing%20and%20modernizing%20the%20country.
Chinese 86/3 Program, 10th revision https://en.wikipedia.org/wiki/863_Program
Chinese Auto Industry Development Policy 2004 https://en.most.gov.cn/newsletters/2004/200411/t20041130_17772.htm
Chinese electric vehicle industry association CEVA http://www.ceva.org.cn/EN/
Chinese National Development and Reform Commission https://en.ndrc.gov.cn/
Chrysler https://www.chrysler.com/
Chrysler 'TEVan' https://en.wikipedia.org/wiki/Chrysler_TEVan
Citroen https://www.citroen.com/en
Citroen 'AX électrique' https://citroenvie.com/citroen-celebrates-20-years-of-pioneering-the-electric-automobile/
Citroen 'Saxo électrique' https://www.citroenet.org.uk/miscellaneous/saxo-electrique/e-saxo.html

Citroën Ami https://www.citroen.co.uk/ami?gclid=2e5e84c5bcf818039060e23da652dc85&gclsrc=3p.ds&msclkid=2e5e84c5bcf818039060e23da652dc85
Citroën ë-Berlingo https://www.citroen.co.uk/models/new-e-berlingo.html
Citroën ë-C4 https://www.citroen.co.uk/models/new-e-c4-x.html?gclid=a3866dbd0aef1574325b0b451169c96a&gclsrc=3p.ds&msclkid=a3866dbd0aef1574325b0b451169c96a
Citroën ë-Jumpy https://en.wikipedia.org/wiki/Citro%C3%ABn_Jumpy
Copilot https://www.microsoft.com/en-us/microsoft-copilot
Coventry, UK (Future Transport Zone) https://www.coventry.ac.uk/research/areas-of-research/centre-for-future-transport-and-cities/
Cruise https://www.getcruise.com/
Cupra Born https://www.cupraofficial.co.uk/new-cars/born/v2
Daimler AG https://group.mercedes-benz.com/en/
Daimler F-Cell https://group.mercedes-benz.com/responsibility/sustainability/climate-environment/environmental-check/glc-f-cell.html
David King, the UK government's former chief scientific adviser on climate change, https://www.gov.uk/government/people/david-king
Deloitte Global Automotive Consumer Study https://www.deloitte.com/global/en/Industries/automotive/perspectives/global-automotive-consumer-study.html
Denza https://www.denzahk.com/en/
Detroit electric car https://en.wikipedia.org/wiki/Detroit_Electric
DG MOVE https://commission.europa.eu/about-european-commission/departments-and-executive-agencies/mobility-and-transport_en
Dialogue Earth https://dialogue.earth/en/climate/introducing-dialogue-earth/
DiDi ride hailing service https://web.didiglobal.com/
Dongfeng Citroën Automobile Company http://www.dongfeng-global.com
Dongfeng EX1 https://carnewschina.com/2022/09/29/dongfeng-ev-ex1-pro-suv-launched-with-price-of-7800-usd/#google_vignette
Dongfeng Fengshen E70 https://carnewschina.com/2022/01/23/dongfeng-e70-launched-with-a-solid-state-battery-50-demonstration-cars-delivered/
DS 3 Crossback E-Tense https://www.dsautomobiles.co.uk/models/ds3.html
DS 4 E-Tense https://www.dsautomobiles.co.uk/models/ds4.html
Eado EV https://www.globalchangan.com/uni.html
Eado PHEV https://www.globalchangan.com/uni.html
E-GMP platform https://www.hyundai.com/worldwide/en/brand-journal/ioniq/e-gmp-revolution
eGT New Energy Automotive Co. Ltd. http://www.dongfeng-global.com/news/details/113
ElectReon https://electreon.com/
ElectReon in Germany (Karlsruhe and Balingen) https://electreon.com/projects/elina-balingen
ElectReon in Sweden (Gotland Island) https://www.smartroadgotland.com/post/successful-test-of-dynamic-charging-of-the-truck#:~:text=ElectReon%20has%20completed%20a%20test%20of%20dynamic%20wireless,the%20Smartroad%20Gotland%20project%20and%20for%20electric%20mobility.
ElectReon in Tel Aviv, Israel https://electreon.com/projects/tel-aviv
electric wheel https://ethw.org/Benjamin_Franklin's_Electric_Motor
Elonroad AB https://www.elonroad.com/
Elväg Syd consortium https://nordicroads.com/evolution-road-new-electric-road-tested-inside/
Emgrand GSe https://global.geely.com/en/models/emgrand
Enfield 465 https://en.wikipedia.org/wiki/Enfield_Automotive
Enfield 8000 ("Electric City Car") https://en.wikipedia.org/wiki/Enfield_Automotive
Enterprise Car Club https://www.enterprisecarclub.co.uk/gb/en/
Equinor https://www.equinor.com/
ERM risks to soil and groundwater https://www.erm.com/solutions/remediation-regeneration/
Ernst & Young https://www.ey.com/en_uk
Étienne Lenoir https://www.britannica.com/biography/Etienne-Lenoir

EU 'Connecting Europe Facility' https://commission.europa.eu/funding-tenders/find-funding/eu-funding-programmes/connecting-europe-facility_en
EU 'Fit for 55' https://www.lr.org/en/services/statutory-compliance/fit-for-55/
EU 'Just Transition Fund' https://commission.europa.eu/funding-tenders/find-funding/eu-funding-programmes/just-transition-fund_en
EU's 'Green Deal' https://commission.europa.eu/strategy-and-policy/priorities-2019-2024/european-green-deal_en
Eulove MPV https://en.wikipedia.org/wiki/Chana_Eulove#Chana_Eulove_EV
Eunice Newton Foote https://www.newscientist.com/article/2382758-eunice-newton-foote-the-woman-who-discovered-the-greenhouse-effect/
European Automobile Manufacturers' Association (ACEA) https://www.acea.auto/
European Commission https://commission.europa.eu/index_en
Everfuel https://www.everfuel.com/
EVolutionRoad https://www.evolutionroad.se/en/news/the-next-generation-electric-road-inaugurated-in-lund-by-southern-swedish-consortium-with-minister-for-energy-and-digital-development-anders-ygeman/
Extended-Range Electric Vehicle' (EREV) powertrain https://energytheory.com/what-are-erevs-extended-range-electric-vehicles/#:~:text=In%20EREVs%20%28Extended-range%20electric%20vehicles%29%2C%20the%20combustion%20engine,generator%20or%20opt%20for%20a%20quick%20DC%20charge.
Fengon E3 https://en.dongfeng-club.com/model/dongfeng-fengon-e3-500-78
Fengshen A60 EV https://www.dfpv.com.cn/
Fertiberia https://www.fertiberia.com/en/
Fiat 500 Electric https://www.fiat.co.uk/models/fiat-500-electric
Fiat Chrysler Automobiles (FCA) https://fcagroup-me.com/en/index.html
Fiat Panda Hydrogen https://www.hydrogencarsnow.com/index.php/fiat-panda-hydrogen-concept/
Fiat Seicento Elettra H2 Fuel Cell https://en.wikipedia.org/wiki/Fiat_Panda_Hydrogen
First Automobile Works (FAW) https://www.faw.com/fawen/gyjt36/fzlc/index.html
Flash Engineering https://www.flashengineering.se/en/about-the-company/flash-engineering/
Ford Escape Hybrid https://www.ford.com/suvs-crossovers/escape/models/escape-plug-in-hybrid/
Ford E-Transit https://www.ford.co.uk/vans-and-pickups/e-transit
Ford Explorer EV https://www.ford.co.uk/cars/electric-explorer#explorer
Ford F-150 Lightning https://www.ford.com/trucks/f150/f150-lightning/2024/
Ford Focus Electric https://en.wikipedia.org/wiki/Ford_Focus_Electric
Ford Mustang Mach-E https://www.ford.co.uk/cars/mustang-mach-e#mache
Ford Mustang Mach-E Extended Range https://www.ford.co.uk/cars/mustang-mach-e#mache
Ford Ranger EV https://www.ford.co.uk/vans-and-pickups/ranger-phev
Ford UK 'the Comuta' https://en.wikipedia.org/wiki/Ford_Comuta
Francois Isaac de Rivaz https://en.wikipedia.org/wiki/Fran%C3%A7ois_Isaac_de_Rivaz
Fraunhofer Institute for Solar Energy Systems ISE https://www.ise.fraunhofer.de/en.html
GAC New Energy Automobile Co. Ltd. https://www.gac.com.cn/en/news/detail?baseid=18636
Galaxy E8 https://www.galaxy-geely.com/E8
Gaston Planté https://www.britannica.com/biography/Gaston-Plante
Geely https://global.geely.com/
Geely EC7-EV https://carnewschina.com/2015/11/19/geely-emgrand-ev-hits-the-chinese-car-market/#google_vignette
General Motors https://www.gm.com/
General Motors 'Impact' www.gm.com/heritage/collection/gm-concept/1990-impact-experimental
General Motors EV1 https://americanhistory.si.edu/explore/exhibitions/ev1-electric-car
Genesis Electrified G80 https://www.genesis.com/worldwide/en/models/luxury-sedan-genesis/electrified-g80/highlights.html
Genesis GV60 https://www.genesis.com/uk/en/models/gv60.html?gclid=CMSVwMeW6IgDFfxmFQgde0M11g&gclsrc=ds

Geometry A https://global.geely.com/en/models/geometrya
Geometry C https://global.geely.com/en/models/geometryc
German Federal Ministry of Transport and Digital Infrastructure https://www.bundesregierung.de/breg-en/federal-government/ministries/federal-ministry-digital-transport
GM "PrEView" https://en.wikipedia.org/wiki/General_Motors_EV1#:~:text=The%20General%20Motors%20EV1%20is%20a%20battery%20electric%20car%20produced#:~:text=The%20General%20Motors%20EV1%20is%20a%20battery%20electric%20car%20produced
GM 'Electrovette' https://en.wikipedia.org/wiki/Chevrolet_Chevette#:~:text=The%20Chevrolet%20Chevette%20was%20a%20subcompact%20hatchback%20produced%20by%20Chevrolet
GM 'Saturn Vue Green Line' https://motorweek.org/road_tests/2007_saturn_vue_green_line/#:~:text=With%20demand%20for%20gasoline-electric%20hybrids%20reaching%20unprecedented%20levels,%20we%20wanted
GM Electrovair https://www.sae.org/news/2023/09/gm-electrovair-history#:~:text=Learn%20how%20GM's%20Electrovair,%20a%20research%20project%20in%201963,%20pioneered
GMC Hummer EV SUV https://www.gmc.com/electric/hummer-ev/suv#:~:text=The%202025%20GMC%20HUMMER%20EV%20SUV%20is%20an%20electric%20vehicle
Goldsworthy Gurney https://www.britannica.com/biography/Goldsworthy-Gurney
Gotion High-Tech https://www.gotion.com/#:~:text=+1-510-249-5610%20info@gotion.com.%2048660%20Kato%20Rd%20Fremont,%20CA%2094538
Green Hydrogen Hub: Eemshaven https://www.rwe.com/en/research-and-development/hydrogen-projects/eemshydrogen/#:~:text=The%20plan%20is%20to%20build%20an%20electrolyser%20for%20the%20production
GRIDSERVE https://www.gridserve.com/
GridX https://www.gridx.ai/
GridX *'Charging Report 2024' https://www.gridx.ai/resources/european-ev-charging-report-2024#:~:text=The%20report%20analyzes%20the%20state%20of%20EV%20charging%20in%20Europe,*
Grumman LLV https://en.wikipedia.org/wiki/Grumman_LLV#:~:text=The%20Grumman%20LLV%20is%20a%20light%20transport%20truck%20designed%20for#:~:text=The%20Grumman%20LLV%20is%20a%20light%20transport%20truck%20designed%20for
Guangzhou Automobile Group Co. Ltd. (GAC Group) https://www.gac-motor.com/en/about/overview.html#:~:text=GAC%20INTERNATIONAL%20is%20the%20international%20business%20division%20of%20Guangzhou%20Automobile
Guangzhou Honda Automobile Co. Ltd. https://www.ghac.cn/#:~:text=%E5%B9%BF%E6%B1%BD%E6%9C%AC%E7%94%B0%E6%98%AF%E5%B9%BF%E6%B1%BD%E9%9B%86%E5%9B%A2%E4%B8%8E
Guangzhou Toyota Motor Co. Ltd. https://www.gac-toyota.com.cn/
Gustave Trouvé https://en.wikipedia.org/wiki/Gustave_Trouv%C3%A9#:~:text=Gustave%20Pierre%20Trouv%C3%A9%20(2%20January%201839%20%E2%80%93%2027%20July%201902)
Guy Stewart Callendar https://www.asme.org/topics-resources/content/the-engineer-who-discovered-global-warming#:~:text=Guy%20Callendar%20was%20a%20combustion%20and%20steam%20engineer%20who%20found
Harrods https://www.harrods.com/
Harrods electric van https://nationalmotormuseum.org.uk/vehicle-collection/harrods-electric-van/#:~:text=Learn%20about%20the%20history%20and%20features%20of%20Harrods'%20electric%20delivery
Henry Ford https://www.britannica.com/money/Henry-Ford https://en.wikipedia.org/wiki/Henry_Ford#:~:text=Henry%20Ford%20was%20the%20founder%20of%20the%20Ford%20Motor%20Company#:~:text=Henry%20Ford%20was%20the%20founder%20of%20the%20Ford%20Motor%20Company

Hippomobile https://en.wikipedia.org/wiki/Hippomobile#:~:text=Hippomobile%20is%20an%20early%20automobile%20with%20an%20internal%20combustion%20engine

Honda https://global.honda/en/#:~:text=Learn%20about%20Honda's%20vision,%20business,%20innovation,%20brand,%20and%20sustainability%20as

Honda 'Insight' https://www.autoexpress.co.uk/honda/insight/359323/honda-insight-mk1-1999-2006-icon-review#:~:text=The%20Honda%20Insight%20was%20the%20first%20mainstream%20hybrid%20car%20to

Honda Civic Hybrid https://www.honda.co.uk/cars/new/civic-hybrid/overview.html#:~:text=The%20new%20Honda%20Civic%20is%20a%20self-charging,%20full%20hybrid%20car

Honda Clarity Electric https://automobiles.honda.com/clarity%20plug-in-hybrid#:~:text=47-mile%20all-electric%20driving-range%20rating.*%20110%20MPGe%20gas-electric%20rating%20and%2044

Honda Clarity Fuel Cell https://automobiles.honda.com/honda-automobiles/home/clarity-fuel-cell#:~:text=Enjoy%20a%20responsible%20drive%20in%20a%20zero-emission%20vehicle.%20The%202021

Honda e https://www.honda.co.uk/cars/new/e-ny1-electric-suv/overview.html#:~:text=The%20Honda%20e:Ny1%20is%20a%20zero-emission%20electric%20SUV%20with%20a

Honda EV Plus (ZA1) https://global.honda/en/heritage/episodes/1988evplus.html#:~:text=Honda%20EV%20Plus%20cruises%20the%20streets%20of%20America.%20This%20advanced

Honda Fit EV https://www.caranddriver.com/reviews/a15123107/2013-honda-fit-ev-electric-first-drive-review/

Honda Group https://global.honda/en/

Honda Prologue https://automobiles.honda.com/prologue#:~:text=The%20Prologue%20is%20the%20first%20all-electric%20SUV%20from%20Honda%20with

Hongqi E-HS3 https://www.hongqi-auto.com/pages/cartype/cartype_hs3.html#:~:text=HONGQI%20HS3%20is%20a%20luxury%20SUV%20with%20sleek%20design,%20intelligent

Hongqi E-HS9 https://www.hongqi-auto.com/pages/cartype/cartype_E-hs9.html#:~:text=EHS9%20is%20a%20new%20energy%20vehicle%20model%20of%20HONGQI%20AUTO

Hongqi E-QM5 https://www.hongqi-auto.com/pages/cartype/cartype_E-qm5.html#:~:text=Hongqi%20E-QM5%20BEV%20is%20a%20smart%20electric%20vehicle%20with%20a

Hydrogen Use in Internal Combustion Engines. US Department of Energy https://www1.eere.energy.gov/hydrogenandfuelcells/tech_validation/pdfs/fcm03r0.pdf#:~:text=Cell%20Engines%20MODULE%203:%20HYDROGEN%20USE%20IN%20INTERNAL%20COMBUSTION%20ENGINES#:~:text=Cell%20Engines%20MODULE%203:%20HYDROGEN%20USE%20IN%20INTERNAL%20COMBUSTION%20ENGINES

Hydrogen Utility (H2U) Eyre Peninsula Gateway Project https://www.energymining.sa.gov.au/industry/hydrogen-and-renewable-energy/hydrogen-in-south-australia/hydrogen-projects-in-south-australia/the-hydrogen-utility-h2u-eyre-peninsula-gateway#:~:text=The%20Eyre%20Peninsula%20Gateway%20Project%20is%20the%20first%20in%20a

Hymotion https://hynet.co.uk/wp-content/uploads/2019/06/15480_CADENT_HYMOTION_PROJECT_REP.pdf#:~:text=This%20HyMotion%20report%20is%20a%20practical%20road%20map%20to%20using

HySE (Hydrogen Small mobility & Engine technology) https://www.hyse-global.com/en/index_en.html#:~:text=Hydrogen%20Small%20mobility%20&%20Engine%20technology%20Association.%20The%20purpose%20of

HySynergy Project https://www.danfoss.com/en/about-danfoss/news/dds/europe-s-largest-electrolysis-plant-under-construction-is-powered-by-danfoss-drives/#:~:text=The%20

HySynergy%20project%20will%20convert%20renewable%20electric%20energy%20for%20 Everfuel%E2%80%99s
Hyundai BlueOn https://www.danfoss.com/en/about-danfoss/news/dds/europe-s-largest-electrolysis-plant-under-construction-is-powered-by-danfoss-drives/#:~:text=The%20 HySynergy%20project%20will%20convert%20renewable%20electric%20energy%20for%20 Everfuel%E2%80%99s
Hyundai Ioniq 5 https://www.hyundai.com/uk/en/models/new-ioniq5.html#:~:text=Discover%20 the%20new%20IONIQ%205,%20a%20stunning%20and%20advanced%20electric
Hyundai Ioniq 6 https://www.hyundai.com/uk/en/models/ioniq6.html#:~:text=The%20 IONIQ%206%20is%20a%20sleek%20and%20spacious%20electric%20saloon
Hyundai Ioniq 6 Long Range https://www.hyundai.news/eu/articles/press-releases/hyundai-2023-ioniq-6-se-long-range-rwd-estimated-driving-range.html
Hyundai Ioniq Electric https://www.hyundai.com/uk/en/models/new-ioniq5.html?gclid=988fcca3 8a48197e2561aa5afb237838&gclsrc=3p.ds&msclkid=988fcca38a48197e2561aa5afb23783 8&utm_source=bing&utm_medium=cpc&utm_campaign=RM%20-%20Models%20-%20 IONIQ%205&utm_term=hyundai%20ioniq%20electric&utm_content=EV
Hyundai Kona Electric https://www.hyundai.com/uk/en/models/kona-electric.html?gclid =b67094f855c910a23299f2012425f831&gclsrc=3p.ds&msclkid=b67094f855c910a2329 9f2012425f831&utm_source=bing&utm_medium=cpc&utm_campaign=RM%20-%20 Models%20-%20KONA%20Electric&utm_term=hyundai%20kona%20Electric%20 auto&utm_content=Automatic
Hyundai Nexo https://www.hyundai.com/uk/en/models/nexo.html#:~:text=Hyundai%20 NEXO%20is%20a%20hydrogen%20fuel%20cell%20SUV%20that%20combines
Hyundai/Kia Group https://www.hyundaimotorgroup.com/main/mainRecommend
Iberdrola https://www.iberdrola.com/home#:~:text=Iberdrola%20is%20a%20Spanish%20 electricity%20company%20with%20more%20than%20180
Iberdrola Fertiberia Project https://www.iberdrola.com/press-room/news/detail/iberdrola-fertiberia-place-spain-forefront-green-hydrogen-europe-with-1-8-billion-investment-projected-2027
Idemitsu Kosan Co. Ltd. (Idemitsu) https://www.idemitsu.com/en/index.html#:~:text=Idemitsu%20 K
Innovation Skåne AB https://innovationskane.com/#:~:text=Vi%20%C3%A4r%20experter%20 p%C3%A5%20att%20leda,%20utveckla%20och%20driva%20innovation.
Intergovernmental Panel on Climate Change (IPCC) https://www.ipcc.ch/#:~:text=The%20 Intergovernmental%20Panel%20on%20Climate%20Change%20(IPCC)%20provides%20 authoritative%20and
International Agency for Research on Cancer (IARC) https://www.iarc.who.int/#:~:text=IARC%20 is%20a%20WHO%20agency%20that%20conducts%20and%20coordinates%20cancer
International Council on Clean Transportation (ICCT) https://theicct.org/#:~:text=The%20 ICCT%20is%20a%20nonprofit%20organization%20that%20provides%20research%20and
ITM Power https://itm-power.com/
J.D. Power https://www.jdpower.com/
J.D. Power's Initial Quality Study https://www.jdpower.com/business/press-releases/2024-us-initial-quality-study-iqs#:~:text=J.D.%20Power%20releases%20its%202024%20U.S.%20 Initial%20Quality%20Study,%20which
Jaguar I-PACE https://www.jaguar.co.uk/jaguar-range/i-pace/index.html
Jaguar Land Rover (JLR) https://www.jaguarlandrover.com/overview
Jamais Contente https://fr.wikipedia.org/wiki/La_Jamais_contente#:~:text=La%20Jamais%20 contente%20est%20une%20voiture%20%C3%A9lectrique%20construite%20par%20Camille
James Hansen https://www.columbia.edu/~jeh1/#:~:text=Learn%20about%20the%20life%20 and%20work%20of%20Dr.%20James%20E.
James Starley tricycle https://en.wikipedia.org/wiki/James_Starley#:~:text=James%20Starley%20 (21%20April%201830%20%E2%80%93%2017%20June%201881)%20[1]

Japan National Railway https://en.wikipedia.org/wiki/Japanese_National_Railways#:~:text=The%20Japanese%20National%20Railways%20(%E6%97%A5%E6%9C%AC%E5%9B%BD%E6%9C%89%E9%89%84%E9%81%93,%20Nihon%20Kokuy%C5%AB%20Tetsud%C5%8D%20or%20Nippon
Japan Railways Group (JRG) https://en.wikipedia.org/wiki/Japan_Railways_Group#:~:text=Learn%20about%20the%20history,%20ownership,%20and%20operations%20of%20the%20JR
Jean-Baptiste Fourier https://www.britannica.com/biography/Joseph-Baron-Fourier
Jeep Avenger EV https://www.jeep.co.uk/jeep-avenger/new-jeep-avenger
Jeep DJ-5E 'Electruck' https://en.wikipedia.org/wiki/Jeep_DJ#:~:text=DJ-5E%20Electruck%20in%20USPS%20livery%20at%20NASA's%20Glenn%20Research%20Center.
John Blenkinsop https://www.britannica.com/biography/John-Blenkinsop
John Tyndall https://www.britannica.com/biography/John-Tyndall
Karl Benz https://www.britannica.com/biography/Karl-Benz
Karl Benz Patent-Motorwagen https://en.wikipedia.org/wiki/Benz_Patent-Motorwagen#:~:text=The%20Benz%20Patent-Motorwagen,%20built%20by%20Karl%20Benz%20in%201885,%20was#:~:text=The%20Benz%20Patent-Motorwagen,%20built%20by%20Karl%20Benz%20in%201885,%20was
Kawasaki https://www.kawasaki.eu/
Kia EV6 https://www.kia.com/uk/new-cars/ev6/
Kia Motors https://www.kia.com/
KiaRayEVhttps://en.wikipedia.org/wiki/Kia_Ray#:~:text=The%20Kia%20Ray%20(Korean:%20%EA%B8%B0%EC%95%84%20%EB%A0%88%EC%9D%B4)%20is%20a%20city%20car#:~:text=The%20Kia%20Ray%20(Korean:%20%EA%B8%B0%EC%95%84%20%EB%A0%88%EC%9D%B4)%20is%20a%20city%20car
Kia Soul EV https://www.kia.com/uk/new-cars/soul-ev/#:~:text=The%20Kia%20Soul%20EV%20is%20a%20compact%20electric%20SUV%20with
Kraftringen Energi AB https://www.kraftringen.se/privat/#:~:text=Kraftringen%20%C3%A4r%20ett%20lokalt%20energibolag%20som%20erbjuder%20elavtal,%20fj%C3%A4rrv%C3%A4rme,%20solparker
Kyoto Protocol (1997) https://unfccc.int/resource/docs/convkp/kpeng.pdf#:~:text=The%20Kyoto%20Protocol%20is%20an%20international%20agreement%20that%20aims%20to
Leroy Somer https://acim.nidec.com/motors/leroy-somer/about-us/leroy-somer#:~:text=Leroy-Somer%20is%20a%20French%20company%20that%20produces%20electric%20motors,%20drives
Lexus RX400 https://en.wikipedia.org/wiki/Lexus_RX#:~:text=The%20Lexus%20RX%20is%20a%20compact%20and%20mid-size%20luxury%20crossover
Lexus RX450 https://www.lexus.co.uk/new-cars/rx/build?path=grades/ac7ecf2e-0f39-4e44-86b2-9db3e1ce6577/3562cd79-6b39-4951-a92d-7f293da2a017/098a3507-a849-442e-96dc-961373979b78
Lexus RZ 450e https://www.lexus.co.uk/new-cars/rz#:~:text=Introducing%20the%20new%20all-electric%20Lexus%20RZ%20450e%20with%20its%20%E2%80%98spindle
Lexus UX 300e https://media.lexus.co.uk/wp-content/uploads/sites/3/pdf/240206-UX-300e-Press-Pack.pdf
Leyden jar https://www.britannica.com/technology/Leyden-jar https://en.wikipedia.org/wiki/Leyden_jar#:~:text=A%20Leyden%20jar%20is%20an%20early%20form%20of%20capacitor%20that
Li Auto L7 https://ir.lixiang.com/news-releases/news-release-details/li-auto-inc-unveils-li-l8-its-six-seat-large-premium-smart-suv#:~:text=Li%20Auto%20Inc.%20unveils%20two%20new%20models%20of%20large%20premium
Li Auto L8 https://ir.lixiang.com/news-releases/news-release-details/li-auto-inc-unveils-li-l8-its-six-seat-large-premium-smart-suv#:~:text=Li%20Auto%20Inc.%20unveils%20two%20new%20models%20of%20large%20premium

Li Auto L9 https://www.lixiang.com/en/L9#li
lithium iron phosphate (LFP)'Blade' battery https://www.byd.com/eu/blog/BYDs-revolutionary-Blade-Battery-all-you-need-to-know.html#:~:text=Another%20unique%20selling%20point%20of%20the%20blade%20battery,excellent%20thermal%20stability%20and%20is%20substantially%20cobalt%20free.
London County Council (LCC) https://en.wikipedia.org/wiki/London_County_Council#:~:text=The%20London%20County%20Council%20(LCC)%20was%20the%20principal%20local%20government
London Transport Museum https://www.ltmuseum.co.uk/
Lucid Air Dream Edition R https://lucidmotors.com/en-sa/air/dream-edition#:~:text=Lucid%20Air%20Dream%20Edition%20comes%20in%20the%20form%20of%20two
Lund municipality / Future by lund https://www.futurebylund.se/lund-id#:~:text=Future%20by%20Lund%20%C3%A4r%20en%20innovationsplattform%20f%C3%B6r%20utveckling%20av%20h%C3%A5llbara
Lund University of Technology https://www.lunduniversity.lu.se/
Magment https://www.magment.co/#:~:text=Magment%20is%20a%20company%20that%20provides%20wireless%20charging%20infrastructure%20based
Maserati Grecale Folgore https://www.maserati.com/global/en/models/grecale-folgore#:~:text=Grecale%20Folgore%20is%20Maserati's%20first%20all-electric%20SUV,%20engineered%20with%20Formula
Matthew Murray https://en.wikipedia.org/wiki/Matthew_Murray#:~:text=Matthew%20Murray%20(1765%20%E2%80%93%2020%20February%201826)%20was%20an%20English#:~:text=Matthew%20Murray%20(1765%20%E2%80%93%2020%20February%201826)%20was%20an%20English
McKinsey https://www.mckinsey.com/
McKinsey & Company https://www.mckinsey.com/
MEB platform https://www.volkswagen-newsroom.com/en/modular-electric-drive-matrix-meb-3677
Mercedes Group https://group.mercedes-benz.com/en/
Mercedes Benz 'A-Class E-Cell' https://en.wikipedia.org/wiki/Mercedes-Benz_A--Class#:~:text=The%20Mercedes-Benz%20A-Class%20E-CELL%20was%20introduced%20in%20September%202010%20and
Mercedes-Benz 'B-Class Electric Drive', B250e https://en.wikipedia.org/wiki/Mercedes-Benz_B--Class#:~:text=The%20B-Class%20is%20a%20subcompact%20executive%20car%20and%20small%20MPV
Mercedes-Benz 'EQA' https://www.mercedes-benz.co.uk/passengercars/models/suv/eqa/overview.html#:~:text=Learn%20about%20the%20EQA,%20the%20latest%20electric%20vehicle%20from%20Mercedes-Benz,
Mercedes-Benz 'EQB' https://www.mercedes-benz.co.uk/passengercars/models/suv/eqb/overview.html#:~:text=Discover%20the%20Mercedes%20EQB:%20price%20|%20technology%20|%20exterior%20|
Mercedes-Benz 'EQE' https://www.mercedes-benz.co.uk/passengercars/models/saloon/eqe/overview.html#:~:text=The%20EQE%20Saloon%20is%20a%20fully%20electric%20car%20with%20a
Mercedes-Benz 'EQS SUV' https://www.mercedes-benz.co.uk/passengercars/models/suv/eqs/overview.html#:~:text=Discover%20the%20Mercedes-Benz%20EQS%20SUV,%20a%20luxurious%207%20seater%20electric
Mercedes-Benz 'EQS' https://www.mercedes-benz.co.uk/passengercars/models/saloon/eqs/overview.html#:~:text=Discover%20the%20highlights%20of%20the%20EQS%20Saloon%20from%20Mercedes-Benz:%20Exterior
Mercedes-Benz 'EQT' https://www.evspecs.org/tech-specs/mercedes-benz/eqt/200#:~:text=Mercedes-Benz%20EQT%20200%20Tech%20Specs%201%20Prices%20Price,Charging%20Times%20Select%20State%20of%20Charge%20%28SOC%29%3A%20

Mercedes-Benz 'eSprinter' https://www.mercedes-benz.co.uk/vans/models/esprinter/panel-van/overview.html#:~:text=Learn%20about%20the%20new%20eSprinter,%20an%20electric%20version%20of%20the

Mercedes-Benz 'eVito' https://www.mercedes-benz.co.uk/vans/models/evito/panel-van/overview.html#:~:text=Learn%20about%20the%20eVito%20Panel%20Van,%20a%20sustainable%20and%20powerful

Mercedes-Benz 'EQC'. The EQC 400 4MATIC https://ev-database.org/uk/car/1337/Mercedes-Benz-EQC-400-4MATIC#:~:text=Mercedes-Benz%20EQC%20400%204MATIC%20December%202020%20-%20June,Battery%20230%20mi%20Real%20Range%20348%20Wh%2Fmi%20Efficiency

Mercedes-Benz EQS 450+ https://www.mercedes-benz.co.uk/passengercars/models/saloon/eqs/overview.html#:~:text=Explore%20the%20features%20and%20specifications%20of%20the%20EQS%20Saloon,%20the

Mercedes-Benz 'EQV' https://www.mercedes-benz.co.uk/passengercars/models/van/eqv/overview.html#:~:text=Discover%20the%20Mercedes-EQ%20EQV:%20MBUX%20|%20Exterior%20|%20Interior%20|

Mercedes-Benz GLS https://www.mercedes-benz.co.uk/passengercars/models/suv/gls/overview.html#:~:text=Discover%20the%20new%20Mercedes-Benz%20GLS,%20a%20luxury%20SUV%20with%20off-road

Mercedes-Benz Group AG https://group.mercedes-benz.com/en/

Mercedes-Benz VISION EQXX https://www.mercedes-benz.com/en/innovation/concept-cars/vision-eqxx-the-new-benchmark-of-effiency/#:~:text=The%20VISION%20EQXX%20is%20a%20prototype%20that%20demonstrates%20how%20Mercedes-Benz

Ministry of Economy, Trade and Industry, Japan (METI) https://www.meti.go.jp/english/#:~:text=METI%20is%20the%20Japanese%20government%20agency%20responsible%20for%20economic%20and

MG Dynamo EV https://www.smmt.co.uk/2014/04/mg-dynamo-launched/#:~:text=MG%20Dynamo%20concept%20EV%20unveiled%20at%20SMMT.%20MG%20is%20celebrating

MG E4 https://www.mg.co.uk/new-cars/mg4-ev#:~:text=MG4%20EV%20is%20a%20new%20electric%20car%20with%20up%20to

MG EZS/ZS EV https://www.mg.co.uk/new-cars/mg-zs-ev#:~:text=MG%20ZS%20EV%20is%20a%20long%20range%20electric%20car%20with

MG5 Long Range https://www.mg.co.uk/new-cars/mg5-ev#:~:text=MG5%20EV%20Long%20Range%20is%20a%20practical,%20affordable%20and%20efficient

Milton O. Reeves and Marshall T. Reeves http://vintagemachinery.org/MfgIndex/detail.aspx?id=1370#:~:text=The%20Reeves%20Pulley%20Co.%20was%20founded%201888%20by%20Marshall%20T.

MINI Cooper SE https://www.mini.co.uk/en_GB/home/range/all-electric-mini-cooper.html#:~:text=Discover%20the%20new%20all-electric%20MINI%20Cooper,%20a%20battery%20electric%20vehicle

MINI E https://www.mini.co.uk/en_GB/home/the-electric-mini-family.html#:~:text=Explore%20the%20new%20all-electric%20MINI%20range,%20including%20Aceman,%20Cooper%20and

Ministry of Industry and Information Technology (MIIT) of China https://english.www.gov.cn/state_council/2014/08/23/content_281474983035940.htm#:~:text=The%20ministry%20is%20a%20State%20Council%20department%20that%20oversees%20China's

Mitsubishi https://www.mitsubishi-motors.com/en/

Mitsubishi Outlander https://mitsubishi-motors.co.uk/model-range/outlander-phev/#:~:text=The%20Outlander%20PHEV%20is%20a%20versatile%20and%20efficient%20SUV%20with

Motor Trend https://www.motortrend.com/

Musée des Arts et Métiers, Paris https://www.arts-et-metiers.net/

NAPAC Fuji SUPER TEC 24 Hours Race https://global.toyota/en/newsroom/corporate/39234866.html#:~:text=Toyota%20Motor%20Corporation%20(Toyota)%20will%20enter%20the%20ENEOS%20Super%20Taikyu
National Road and Transport Research Institute, Sweden https://www.vti.se/en/
NCM chemistry batteries https://www.electricity-magnetism.org/electric-battery/nmc-battery/
NEOM Green Hydrogen Project https://www.neom.com/en-us/newsroom/neom-hydrogen-project
Netherlands Organization for Applied Scientific Research (TNO) https://www.tno.nl/en/about-tno/organisation/#:~:text=TNO%20is%20an%20independent%20not-for-profit%20research%20organisation%20that%20applies%20knowledge
New York State 'Electronic Equipment Recycling and Reuse Act' https://dec.ny.gov/environmental-protection/recycling-composting/electronic-waste-recycling
New York Times https://www.nytimes.com/
Nicolas-Joseph Cugnot https://www.britannica.com/biography/Nicolas-Joseph-Cugnot
Nikolaus Otto https://www.britannica.com/biography/Nikolaus-Otto
NIO https://www.nio.com
Nissan https://www.nissan-global.com/EN/
Nissan 'Altra EV' https://www.nissan-global.com/EN/HERITAGE_COLLECTION/451_rnessa_ev.html#:~:text=Nissan%20has%20continued%20R&D%20of%20EV%20that%20originated%20from%20%22Tama
Nissan Ariya https://www.nissan.co.uk/vehicles/new-vehicles/ariya/price-specifications.html#:~:text=Nissan%20ARIYA%20Prices%20&%20Specifications.%2063kWh%20ENGAGE.%20Prices%20Range%20from.
Nissan Avenir EV in 1994 https://www.nissan-global.com/EN/HERITAGE_COLLECTION/451_rnessa_ev.html#:~:text=Nissan%20has%20continued%20R&D%20of%20EV%20that%20originated%20from%20%22Tama
Nissan Cedric EV https://www.nissan-global.com/EN/HERITAGE_COLLECTION/451_rnessa_ev.html#:~:text=Nissan%20has%20continued%20R&D%20of%20EV%20that%20originated%20from%20%22Tama
Nissan e-NV200 https://www.drivingelectric.com/nissan/e-nv200/range#:~:text=The%20Nissan%20e-NV200%20is%20an%20electric%20van%20with%20a%2040kWh
Nissan Leaf https://en.wikipedia.org/wiki/Nissan_Leaf#:~:text=The%20Nissan%20Leaf%20is%20a%20battery-electric%20car%20produced%20since%202010
Nissan Prairie Joy EV https://www.nissan-global.com/EN/HERITAGE_COLLECTION/450_prairie_joy_ev.html#:~:text=The%20lease%20of%20this%20%E2%80%9CPrairie%20Joy%20EV,%E2%80%9D%20the%20world's%20first
Nissan Qashqai ePower https://www.nissan.co.uk/vehicles/new-vehicles/qashqai.html?cid=psm_cmid=506129173_grid=1260041971113632_adid=&&msclkid=ea9d88ff5acd1af2478eee020606770a&gclid=ea9d88ff5acd1af2478eee020606770a&gclsrc=3p.ds
Nissan Sakura https://www.nissan-global.com/EN/INNOVATION/DESIGN/DESIGNWORKS/SAKURA/#:~:text=Following%20the%20Nissan%20LEAF%20and%20Ariya,%20the%20all-new,%20100%%20electric
Nissan Townstar EV https://www.nissan.co.uk/vehicles/new-vehicles/townstar/electric-van.html#:~:text=The%20Nissan%20Townstar%20is%20a%20clean,%20quiet%20and%20cost-effective%20electric
Nissan X-Trail ePower https://www.nissan.co.uk/vehicles/new-vehicles/x-trail.html#:~:text=Explore%20the%20Nissan%20X-Trail,%20a%207-seater%20car%20with%20e-POWER%20technology
NorthH2 Project (Netherlands) https://www.rwe.com/en/research-and-development/hydrogen-projects/north2/#:~:text=NortH2%20is%20a%20project%20by%20RWE%20and%20other%20partners%20to
Opel Ampera https://www.media.stellantis.com/em-en/opel/press/happy-tenth-birthday-ampera-opel-s-pioneering-electric-car

Opel/Vauxhall Combo-e Life https://www.vauxhall.co.uk/cars/combo-life/model-overview.html#:~:text=Combo%20Life%20Electric%20is%20a%20spacious%20and%20versatile%20family%20car

Opel/Vauxhall Corsa-e https://www.vauxhall.co.uk/cars/new-corsa/electric.html#:~:text=Explore%20the%20new%20Corsa%20Electric,%20a%20pure%20electric%20car%20with

Opel/Vauxhall Mokka-e https://www.vauxhall.co.uk/cars/mokka/electric.html#:~:text=Discover%20the%20Vauxhall%20Mokka%20Electric,%20a%20100%%20electric%20car%20with

Opel/Vauxhall Vivaro-e https://www.vauxhall.co.uk/cars/vivaro-life/model-overview.html#:~:text=The%20Vivaro%20Life%20Electric%20is%20a%20fully%20electric%20version%20of

OpenAI https://openai.com/

Ørsted https://orsted.com/

Peugeot https://www.peugeot.com/en/

Peugeot e-2008 https://www.peugeot.co.uk/models/new-peugeot-2008/electric.html

Peugeot e-208 https://offers.peugeot.co.uk/choose-your-peugeot/configure-208-and-e-208 https://www.peugeot.co.uk/about-us/brand/news-articles/all-new-peugeot-e-208-wins-best-electric-car.html

Peugeot e-Expert https://www.peugeot.co.uk/models/peugeot-expert.html

Pierre Faure https://www.reddit.com/r/WeirdWheels/comments/fo30ki/pierre_faur%C3%A9s_electra_an_electric_city_car_from/#:~:text=This%20was%20one%20of%20the%20many%20electric%20cars%20which%20were

Plugless Power https://www.pluglesspower.com/

Polestar https://www.polestarglobal.com/

Polestar 1 https://www.polestar.com/uk/polestar-1/#:~:text=Polestar%201%20is%20a%20limited%20edition%20hybrid%20GT%20car%20with

Polestar 2 https://www.polestar.com/uk/polestar-2/

Polestar 3 https://www.polestar.com/uk/polestar-3/

Polestar 5 https://www.polestar.com/global/precept/#:~:text=A%20vision%20that%20will%20be%20realised%20as%20Polestar%205,%20set

Porsche https://www.porsche.com/international/

Porsche Taycan https://www.porsche.com/uk/models/taycan/#:~:text=Find%20out%20the%20features,%20prices%20and%20specifications%20of%20the%20Taycan

Porsche Taycan Cross Turismo https://www.porsche.com/uk/models/taycan/taycan-cross-turismo-models/taycan-turbo-cross-turismo/

PSA Group (Peugeot Société Anonyme) https://www.peugeot.fr/marque/univers-peugeot/

PSA Suzuki https://www.autocarpro.in/news-international/psa-suzuki-set-ev-business-divisions-28954

Purdue University and the Indiana Department of Transportation (INDOT) https://www.purdue.edu/newsroom/2024/Q2/building-the-first-highway-segment-in-the-u-s-that-can-charge-electric-vehicles-big-and-small-as-they-drive/#:~:text=At%20the%20%E2%80%9CCrossroads%20of%20America,%E2%80%9D%20Purdue%20University%20engineers%20and%20the

Qinchuan https://www.qinchuan.com/en/#:~:text=Shaanxi%20Qinchuan%20Grinding%20Machine%20Co.,%20Ltd.%20Qinchuan%20Group%20Baoji%20Instrument

Qualcomm Halo https://www.qualcomm.com/news/onq/2017/05/wireless-dynamic-ev-charging-evolution-qualcomm-halo#:~:text=Qualcomm%20demonstrates%20the%20attributes%20of%20Qualcomm%20Halo%20wireless%20electric%20vehicle

QuantumScape https://www.quantumscape.com/

Ramboll AB https://www.ramboll.com/

REFHYNE Project https://www.refhyne.eu/#:~:text=REFHYNE%20is%20a%20first-of-a-kind%20project%20at%20the%20forefront%20of%20the

Renault Dauphine https://en.wikipedia.org/wiki/Renault_Dauphine#:~:text=The%20Renault%20Dauphine%20(pronounced%20[d%C9%94fin])%20[10]%20is%20an%20economy%20car#:~:text=The%20Renault%20Dauphine%20(pronounced%20[d%C9%94fin])%20[10]%20is%20an%20economy%20car

Research Institute of Electric Machines in Brno, EMA1 https://en.brno.cz/w/scientific-research-centres-in-brno#:~:text=Brno%20is%20one%20of%20the%20most%20progressive%20centres%20of%20science

Reuters https://www.reuters.com/

Richard Trevithick https://en.wikipedia.org/wiki/Richard_Trevithick#:~:text=Richard%20Trevithick%20(13%20April%201771%20%E2%80%93%2022%20April%201833)%20was

Robert Anderson https://blog.upsbatterycenter.com/robert-anderson-19th-century-scottish-inventor/#:~:text=Robert%20Anderson%20is%20known%20as%20the%20inventor%20of%20the%20first

Roewe E50 https://en.wikipedia.org/wiki/Roewe_E50

Roewe Marvel R https://www.chinapev.com/saic/roewe/roewe-debuted-its-first-5g-ev-marvel-r/#:~:text=At%20the%20Roewe%20Brand%20Day%20on%20May%2010,%20Roewe%20released

Rudolf Diesel https://www.britannica.com/biography/Rudolf-Diesel

SAIC https://www.saicmotor.com/english/

Samsung https://www.samsung.com/global/common/samsung/

Samuel's Electric Carriage and Wagon Company https://virtualtechnologymuseum.com/s/transportation/page/firstelectrictaxi#:~:text=Samuel%E2%80%99s%20Electric%20Carriage%20and%20Wagon%20Company%20introduced%20its%20own%20fleet https://en.wikipedia.org/wiki/Electric_Vehicle_Company#:~:text=In%20May%201897,%20Rice%20had%20acquired%20the%20Electric%20Carriage%20&

Scottish Aviation's 'Scamp' https://en.wikipedia.org/wiki/Scottish_Aviation_Scamp#:~:text=The%20Scottish%20Aviation%20Scamp%20is%20a%20small%20concept%20electric%20city#:~:text=The%20Scottish%20Aviation%20Scamp%20is%20a%20small%20concept%20electric%20city

SEAT Mii Electric https://www.seat.co.uk/new-cars/mii-electric/overview#:~:text=The%20new%20SEAT%20electric%20car%20is%20designed%20to%20be%20the

SEAT/Cupra https://www.seat-mediacenter.com/ https://CUPRAofficial.com

Sebring-Vanguard 'CitiCar' https://en.wikipedia.org/wiki/Citicar#:~:text=The%20CitiCar%20is%20a%20car%20produced%20from%201974%20to%201977#:~:text=The%20CitiCar%20is%20a%20car%20produced%20from%201974%20to%201977

Shanghai Volkswagen Automotive Co Ltd https://volkswagengroupchina.com.cn/en/volkswagengroupchina/

Shell https://www.shell.com/

Shenzhen BYD Daimler New Technology' https://carnewschina.com/2024/09/16/mercedes-benz-out-byd-takes-100-ownership-of-denza-in-china/#google_vignette

Sibrandus Stratingh https://www.rug.nl/museum/collections/collection-stories/wagentje-van-stratingh?lang=en#:~:text=The%20car%20was%20designed%20by%20professor%20of%20chemistry%20and%20technology

Siemens https://www.siemens.com/global/en.html

Sila Nanotechnologies https://www.silanano.com/#:~:text=Innovative%20materials%20by%20Sila%20Nanotechnologies%20for%20fast%20EV%20charging%20&

Sinclair C5 https://en.wikipedia.org/wiki/Sinclair_C5#:~:text=The%20Sinclair%20C5%20is%20a%20small%20one-person%20battery%20electric%20recumbent

SiTech MEV 100 https://carnewschina.com/2018/04/20/sitech-is-a-new-ev-brand-from-china-and-this-is-their-first-car/#google_vignette

Skånetrafiken AB https://www.skanetrafiken.se/#:~:text=Res%20h%C3%A5llbart%20och%20med%20l%C3%A4tthet%20runt%20i%20Sk%C3%A5ne%20med%20v%C3%A5ra

Skeleton Technologies https://www.skeletontech.com/#:~:text=Skeleton%20Technologies%20is%20the%20world's%20leading%20manufacturer%20of%20graphene-based%20supercapacitors.
Škoda https://www.skoda-auto.com/
Škoda Citigo-e Iv https://www.skoda-auto.com/_doc/702b718e-8753-45e8-a272-b724c2ee0832#:~:text=%C5%A0KODA%20has%20unveiled%20the%20first%20fully%20electric%20car.%20The%20power
Škoda Enyaq iV https://www.skoda-auto.com/models/range/enyaq
Smart Fortwo Electric Drive https://www.smart.mercedes-benz.com/int/en/home
Society of Motor Manufacturers and Traders https://www.smmt.co.uk/
Solaris Sverige AB https://www.solarisbus.com/se
Solid Power https://www.solidpowerbattery.com/all-solid-state-batteries/default.aspx#:~:text=Solid%20Power%E2%80%99s%20all-solid-state%20battery%20cell%20technology%20is%20expected%20to%20provide
Sony Corporation https://www.sony.net/
State-owned Assets Supervision and Administration Commission of Beijing Municipal Government. https://english.beijing.gov.cn/government/departments/202006/t20200625_1932642.html#:~:text=To%20implement%20laws,%20rules,%20regulations%20and%20policies%20regarding%20state-owned%20assets
Stellantis and Arena del Futuro (Italy) https://www.stellantis.com/en/news/press-releases/2021/december/arena-del-futuro-innovative-dynamic-induction-charging-becomes-a-reality#:~:text=TURIN%20%E2%80%93%20After%20years%20of%20intense,%20in-depth%20studies,%20%E2%80%9CArena%20Del
Stellantis N.V. https://www.stellantis.com/en
Stephenson's https://en.wikipedia.org/wiki/George_Stephenson#:~:text=George%20next%20paid%20attention%20to%20Anne%20Henderson%20where%20he%20lodged
Subaru https://www.subaru.co.jp/en/#:~:text=The%20technology%20that%20makes%20Subaru%20different:%20enjoyment%20and%20peace%20of
Svante Arrhenius https://www.britannica.com/biography/Svante-Arrhenius
Swedish Transport Administration https://www.government.se/government-agencies/swedish-transport-administration/
Sylvester Roper https://en.wikipedia.org/wiki/Sylvester_H._Roper#:~:text=Sylvester%20Howard%20Roper%20(November%2024,%201823%20%E2%80%93%20June%201,%201896)
T3 ride hailing service https://www.smartshanghai.com/articles/activities/tested-t3-the-new-didi-alternative-that-let-s-you-ride-like-a-diplomat#:~:text=T3%20doesn%E2%80%99t%20have%20an%20English-language%20option%20yet%20but%20the%20interface
Tama https://en.wikipedia.org/?title=Tama_Electric_Car_Company&redirect=no#:~:text=Retrieved%20from
Technode.com https://technode.com/
TEHA business forum in Cernobbio https://www.ambrosetti.eu/en/our-story/cernobbio-forum-50-years/#:~:text=%EF%BB%BFThe%20Cernobbio%20Forum%20reflects%20TEHA%E2%80%99s%20claim,%20%22the%20future%20today%22,%20because
Tesla Cybertruck https://www.tesla.com/cybertruck#:~:text=View%20Cybertruck%20Off-Road%20Guide.%20Built%20for%20any%20adventure,%20Cybertruck%20has
Tesla Model 3 https://www.tesla.com/en_gb/model3#:~:text=Model%203%20|%20Tesla%20United%20Kingdom.%20Built%20for%20Distance.%20Go
Tesla Model S https://www.tesla.com/en_gb/models#:~:text=Model%20S%20platforms%20unite%20powertrain%20and%20battery%20technologies%20for%20unrivaled
Tesla Model S Long Range https://www.autoexpress.co.uk/tesla/model-s/107620/new-tesla-model-s-long-range-2019-review#:~:text=The%20all-electric%20Tesla%20Model%20S%20has%20been%20updated%20for%202019,

Tesla Model X https://www.tesla.com/modelx#:~:text=Model%20X%20is%20built%20for%20utility%20and%20performance,%20with%20standard
Tesla Model Y https://www.tesla.com/en_gb/modely#:~:text=Model%20Y%20is%20a%20fully%20electric,%20mid-size%20SUV%20with%20unparalleled
Tesla Motors https://www.tesla.com/#:~:text=Tesla%20is%20accelerating%20the%20world's%20transition%20to%20sustainable%20energy%20with
Tesla Roadster https://www.tesla.com/roadster#:~:text=An%20all-electric%20vehicle,%20the%20Tesla%20Roadster%20is%20the%20quickest%20car
Tesla Semi https://www.tesla.com/semi#:~:text=Tesla%20Semi%20is%20a%20fully%20electric%20semi%20truck%20with%20active
The Economist https://www.economist.com/
the Electric Vehicle Company https://en.wikipedia.org/wiki/Electric_Vehicle_Company#:~:text=The%20company's%20arm%20in%20Washington,%20D.C.,%20had%20a%20circular%20showroom
The Henney Kilowatt https://en.wikipedia.org/wiki/Henney_Kilowatt#:~:text=The%20Henney%20Kilowatt%20was%20an%20electric%20car%20introduced%20in%20the
The Register https://www.theregister.com/#:~:text=Microsoft%20offers%20a%20way%20out.%20Might%20be%20best%20to%20give
The Standard Electric https://en.wikipedia.org/wiki/Standard_Electric_(automobile)#:~:text=History.%20The%20Standard%20Electric%20used%20electric%20motors%20made%20by%20Westinghouse
The Telegraph https://www.telegraph.co.uk/
The Urban Electric Network https://www.urbanelectric.london/#:~:text=Urban%20Electric%20enables%20you%20to%20conveniently%20charge%20an%20electric%20vehicle
Thomas Wilson https://en.wikipedia.org/wiki/William_Murdoch#:~:text=The%20earliest%20mention%20of%20Murdoch's%20thoughts%20and%20plans%20for%20this#:~:text=The%20earliest%20mention%20of%20Murdoch's%20thoughts%20and%20plans%20for%20this
Timothy Hackworth https://en.wikipedia.org/wiki/Timothy_Hackworth#:~:text=Timothy%20Hackworth%20(22%20December%201786%20%E2%80%93%207%20July%201850)%20was#:~:text=Timothy%20Hackworth%20(22%20December%201786%20%E2%80%93%207%20July%201850)%20was
TNO (Netherlands Organization for Applied Scientific Research) https://www.tno.nl/en/about-tno/organisation/#:~:text=TNO%20is%20an%20independent%20not-for-profit%20research%20organisation%20made%20up%20of
Tokyo Electro Automobile Co. Tama https://www.nissan-global.com/EN/HERITAGE_COLLECTION/tama_electric.html#:~:text=%22TAMA%22%20is%20an%20electric%20vehicle%20developed%20by%20Tokyo%20Electro%20Automobile
Tongji University https://www.tongji.edu.cn/eng/#:~:text=16%20innovation%20achievements%20of%20Tongji%20University%20appear%20at%20the%2023rd
Total https://totalenergies.com/
TotalEnergies https://totalenergies.com/#:~:text=TotalEnergies%20is%20a%20global%20integrated%20energy%20company%20that%20produces%20and
Toyota https://global.toyota/en/
Toyota bZ3 https://global.toyota/en/newsroom/toyota/38131700.html#:~:text=Toyota%20Motor%20Corporation%20(Toyota)%20has%20announced%20the%20Toyota%20bZ3,%20the
Toyota bZ4X https://www.toyota.co.uk/new-cars/bz4x#:~:text=Toyota's%20All-Electric%20bZ4X%20gives%20you%20the%20freedom%20to%20go%20where
Toyota Mirai https://www.toyota.co.uk/new-cars/mirai#:~:text=The%20Mirai%20is%20the%20next%20step%20in%20the%20age%20of
Toyota Media https://media.toyota.co.uk/toyota-and-woven-planet-develop-new-portable-hydrogen-cartridge-prototype-to-power-everyday-applications/

Toyota Newsroom https://newsroom.toyota.eu/
Toyota Prius https://en.wikipedia.org/wiki/Toyota_Prius#:~:text=The%20Toyota%20Prius%20(/%20%CB%88%20p%20r%20i%CB%90%20%C9%99%20s https://www.toyota.com/prius/
Toyota Prius Prime https://www.toyota.com/priusprime/
Toyota RAV4 EV https://en.wikipedia.org/wiki/Toyota_RAV4_EV#:~:text=The%20Toyota%20RAV4%20EV%20is%20an%20all-electric%20version%20of%20the
Toyota RAV4 EV (Second Generation) https://en.wikipedia.org/wiki/Toyota_RAV4_EV#:~:text=The%20Toyota%20RAV4%20EV%20is%20an%20all-electric%20version%20of%20the https://www.toyota.co.uk/new-cars/rav4-plugin#:~:text=The%20RAV4%20Plug-in%20Hybrid%20is%20responsive%20and%20delivers%20an%20enjoyable
Transport for London https://tfl.gov.uk/
Tsinghua University https://www.tsinghua.edu.cn/en/
UK's Future Transport Zones program https://www.gov.uk/government/collections/future-of-transport-programme#:~:text=Information%20on%20the%20Future%20of%20Transport%20programme,%20which%20aims%20to
UNECE Categories M1/N1 https://unece.org/transport/vehicle-regulations/classification-and-definition-vehicles#:~:text=UNECE;%20Transport;%20Vehicle%20Regulations;%20Classification%20and%20Definition%20of%20Vehicles%20.
Unique Mobility 'Electrek' https://www.uniquemobility.org/history#:~:text=In%201979%20Unique%20Mobility%20unveiled%20its%20first-ever%20electric,the%20first%20commercial%20vehicles%20to%20include%20regenerative%20braking.
United Nations https://www.un.org/en/
United Nations Framework Convention on Climate Change (UNFCCC) https://unfccc.int/process-and-meetings/what-is-the-united-nations-framework-convention-on-climate-change#:~:text=Directs%20new%20funds%20to%20climate%20change%20activities%20in%20developing%20countries.
USA Department of Energy's Hydrogen Shot Initiative https://www.energy.gov/eere/fuelcells/hydrogen-shot
VAZ 2801 https://www.rbth.com/history/334125-electric-cars-made-in-ussr-soviet#:~:text=VAZ-2801.%20Vitaly%20Sozinov/TASS.%20This%20electric%20car%20was%20made%20at%20the
Venucia e30 https://global.nissannews.com/en/releases/release-a10789d584980a1255912f21073dc0b1-140910-04-e#:~:text=SHANGHAI,%20China%20-%20Dongfeng%20Nissan%20Passenger%20Vehicle%20Company%20(DFL-PV)%20today
Volkswagen / Volkswagen Group https://www.volkswagen-group.com/en
Volkswagen e-Golf https://www.volkswagen.co.uk/en/new/e-golf.html#:~:text=Find%20out%20more%20about%20the%20new%20e-Golf,%20an%20electric%20car
Volkswagen e-Up! https://www.volkswagen.co.uk/en/new/e-up.html#:~:text=Explore%20the%20Volkswagen%20e-up!%20Add%20your%20personality,%20comfort%20and%20style
Volkswagen ID. Buzz https://www.volkswagen.co.uk/en/electric-and-hybrid/electric-cars/id-buzz.html#:~:text=It%20is%20longer%20and%20can%20go%20further:%20thanks%20to%20the#:~:text=It%20is%20longer%20and%20can%20go%20further:%20thanks%20to%20the
Volkswagen ID.3 https://www.volkswagen.co.uk/en/electric-and-hybrid/electric-cars/id3.html#:~:text=Built%20on%20the%20modular%20electric%20drive%20matrix%20(MEB),%20the%20ID.3
Volkswagen ID.4 https://www.volkswagen.co.uk/en/electric-and-hybrid/electric-cars/id4.html#:~:text=Explore%20the%20new%20Volkswagen%20ID.4.%20The%20ID.4%20gives%20you%20the
Volkswagen ID.5 https://www.volkswagen.co.uk/en/electric-and-hybrid/electric-cars/id5-gtx.html?utm_source=bing&utm_medium=ppc&utm_campaign=promotion&campaign=00812940|Brand-Model-ID4&language=EN&adchan=seah&publisher=GADW&country=GB&adplt=pseac&cpid=71700000115281201|58700008576408423&adver=new+volkswagen+id5+gtx&format=NU&adpay=pd&adcta=NU&adaud=NU&adpl=GADW&adcr=new+volkswagen+id5

+gtx&adgroup=Trim_GTX&seatype=branded&gclid=734174f67b5b186bdbee13f753bba885&gclsrc=3p.ds&msclkid=734174f67b5b186bdbee13f753bba885

Volkswagen ID.7 https://www.volkswagen.co.uk/en/electric-and-hybrid/electric-cars/id7.html?campaign=00812940|Brand-Model-ID7&language=EN&adchan=seah&publisher=GADW&country=GB&adplt=pseac&cpid=71700000115281138|58700008575898526&adver=volkswagen+id+7&format=NU&adpay=pd&adcta=NU&adaud=NU&adpl=GADW&adcr=volkswagen+id+7&adgroup=ID7_Core_MS&seatype=branded&gclid=7d8c724698c01492c2d51547b18b1e2f&gclsrc=3p.ds&msclkid=7d8c724698c01492c2d51547b18b1e2f

Volvo https://www.volvocars.com/

Volvo 'C40 Recharge' https://www.volvocars.com/uk/cars/ec40-electric/

Volvo C30 https://www.media.volvocars.com/global/en-gb/media/pressreleases/17402#:~:text=Launched%20in%202006,%20the%20Volvo%20C30%20is%20based%20on%20the

Volvo EX90 https://www.volvocars.com/uk/cars/ex90-electric/

Volvo SCC (Safety Concept Car) https://www.media.volvocars.com/global/en-gb/models/scc/0#:~:text=The%20main%20theme%20of%20the%20Volvo%20Safety%20Concept%20Car%20(SCC)

Volvo XC40 Recharge https://www.volvocars.com/uk/cars/xc40-electric/specification/

VTI https://www.vti.se/en/#:~:text=We%20are%20located%20in%20Sweden%20with%20offices%20in%20Link%C3%B6ping,%20Stockholm,

Walter Bersey https://en.wikipedia.org/wiki/Walter_Bersey#:~:text=Walter%20Charles%20Bersey%20(15%20October%201874%20%E2%80%93%2021%20April%201950)

Wang Chuanfu https://en.wikipedia.org/wiki/Wang_Chuanfu#:~:text=Wang%20Chuanfu%20(Chinese:%20%E7%8E%8B%E4%BC%A0%E7%A6%8F;%20born%208%20April%201966)%20is%20a#:~:text=Wang%20Chuanfu%20(Chinese:%-20%E7%8E%8B%E4%BC%A0%E7%A6%8F;%20born%208%20April%201966)%20is%20a

Wards 100 https://wardsauto.informa.com/wards100/#:~:text=As%20an%20organization,%20we%20remain%20committed%20to%20carrying%20on%20Al

Waymo https://waymo.com/#:~:text=Waymo%E2%80%94formerly%20the%20Google%20self-driving%20car%20project%E2%80%94makes%20it%20safe%20and%20easy

WEEE Directive https://environment.ec.europa.eu/topics/waste-and-recycling/waste-electrical-and-electronic-equipment-weee_en#:~:text=The%20WEEE%20Directive%20aims%20to%20prevent,%20reuse,%20recycle%20and%20recover

Westinghouse https://westinghouse.com/

Wikipedia https://www.wikipedia.org/#:~:text=Wikipedia%20is%20a%20free%20online%20encyclopedia,%20created%20and%20edited%20by

William Hedley https://www.britannica.com/biography/William-Hedley

William Morrison of Des Moines https://guides.loc.gov/chronicling-america-early-electric-cars#:~:text=William%20Morrison%20from%20Des%20Moines,%20Iowa%20develops%20a%20six-passenger%20electric

William Murdoch https://en.wikipedia.org/wiki/William_Murdoch#:~:text=William%20Murdoch%20(sometimes%20spelled%20Murdock)%20(21%20August%201754%20%E2%80%93%2015

WiTricity https://witricity.com/#:~:text=WiTricity%20has%20expanded%20wireless%20EV%20charging%20to%20golf%20carts,%20low-speed

Yamaha https://www.yamaha.com/en/

Zagato 'Zele' (known in USA as the 'Elcar') https://en.wikipedia.org/wiki/Zagato_Zele#:~:text=The%20Zagato%20Zele%201000,%201500,%202000%20(sold%20in%20the%20United

Zeekr 001 https://www.zeekr.eu/models/001

Zénobe Gramme https://www.britannica.com/biography/Zenobe-Theophile-Gramme

https://www.gov.uk/government/news/pathway-for-zero-emission-vehicle-transition-by-2035-becomes-law#:~:text=The%20UK%20government%20has%20passed%20new%20regulations%20to%20require%20car

Zhejiang Geely Holding Group Co. Ltd (Geely) https://zgh.com/?lang=en#:~:text=Zhejiang%20Geely%20Holding%20Group%20(ZGH)%20is%20the%20lead%20shareholder%20of

Zipcar https://www.zipcar.com/en-gb#:~:text=Zipcar%20is%20a%20car%20sharing%20service%20that%20lets%20you%20rent

ZipCharge https://zipcharge.editorx.io/website

Index

B. Williams, *Electric Shocks: Are Drivers Rejecting the Electric Car?*, https://doi.org/10.1007/978-3-031-98820-2

MIX
Papier aus verantwortungsvollen Quellen
Paper from responsible sources
FSC® C105338

If you have any concerns about our products,
you can contact us on
ProductSafety@springernature.com

In case Publisher is established outside the EU,
the EU authorized representative is:
Springer Nature Customer Service Center GmbH
Europaplatz 3, 69115 Heidelberg, Germany

Printed by Libri Plureos GmbH
in Hamburg, Germany